PROTEINS FROM HYDROCARBONS

PROTEINS FROM HYDROCARBONS

Sidney Gutcho

NOYES DATA CORPORATION
Park Ridge, New Jersey London, England
1973

Library of Congress Catalog Card Number: 72-82913
ISBN: 0-8155-0480-2
Printed in the United States

Published in the United States of America by
Noyes Data Corporation
Noyes Building, Park Ridge, New Jersey 07656

Foreword

The detailed, descriptive information in this book is based on U.S. patents since 1965 relating to the production of proteins from hydrocarbons. In order to round out the complete technological picture, one earlier but very relevant patent was included.

This book serves a double purpose in that it supplies detailed technical information and can be used as a guide to the U.S. patent literature in this field. By indicating all the information that is significant, and eliminating legal jargon and juristic phraseology, this book presents an advanced, commercially oriented review of obtaining proteins from hydrocarbons.

The U.S. patent literature is the largest and most comprehensive collection of technical information in the world. There is more practical, commercial, timely process information assembled here than is available from any other source. The technical information obtained from a patent is extremely reliable and comprehensive; sufficient information must be included to avoid rejection for "insufficient disclosure".

The patent literature covers a substantial amount of information not available in the journal literature. The patent literature is a prime source of basic commercially useful information. This information is overlooked by those who rely primarily on the periodical journal literature. It is realized that there is a lag between a patent application on a new process development and the granting of a patent, but it is felt that this may roughly parallel or even anticipate the lag in putting that development into commercial practice.

Many of these patents are being utilized commercially. Whether used or not, they offer opportunities for technological transfer. Also, a major purpose of this book is to describe the number of technical possibilities available, which may open up profitable areas of research and development. One should have to go no further than this condensed information to establish a sound background before launching into research in this field.

Advanced composition and production methods developed by Noyes Data are employed to bring these durably bound books to you in a minimum of time. Specialized techniques are used to close the gap between "manuscript" and "completed book." Industrial technology is progressing so rapidly that time-honored, conventional typesetting, printing, binding and shipping methods can render a technical or scientific book obsolete before the potential user sees it.

The Table of Contents is organized in such a way as to serve as a subject index. Other indexes by company, inventor and patent number help in providing easy access to the information contained in this book.

15 Reasons Why the U.S. Patent Office Literature Is Important to You-

(1) The U.S. patent literature is the largest and most comprehensive collection of technical information in the world. There is more practical commercial process information assembled here than is available from any other source.

(2) The technical information obtained from the patent literature is extremely comprehensive; sufficient information must be included to avoid rejection for "insufficient disclosure."

(3) The patent literature is a prime source of basic commercially utilizable information. This information is overlooked by those who rely primarily on the periodical journal literature.

(4) An important feature of the patent literature is that it can serve to avoid duplication of research and development.

(5) Patents, unlike periodical literature, are bound by definition to contain new information, data and ideas.

(6) It can serve as a source of new ideas in a different but related field, and may be outside the patent protection offered the original invention.

(7) Since claims are narrowly defined, much valuable information is included that may be outside the legal protection afforded by the claims.

(8) Patents discuss the difficulties associated with previous research, development or production techniques, and offer a specific method of overcoming problems. This gives clues to current process information that has not been published in periodicals or books.

(9) Can aid in process design by providing a selection of alternate techniques. A powerful research and engineering tool.

(10) Obtain licenses - many U.S. chemical patents have not been developed commercially.

(11) Patents provide an excellent starting point for the next investigator.

(12) Frequently, innovations derived from research are first disclosed in the patent literature, prior to coverage in the periodical literature.

(13) Patents offer a most valuable method of keeping abreast of latest technologies, serving an individual's own "current awareness" program.

(14) Copies of U.S. patents are easily obtained from the U.S. Patent Office at 50¢ a copy.

(15) It is a creative source of ideas for those with imagination.

Contents and Subject Index

INTRODUCTION 1

LIQUID HYDROCARBON FEEDSTOCKS 4
- Paraffinic Feedstocks 4
  - Straight Chain C_{10} Paraffinic Hydrocarbons 4
  - Heavy Gas Oils 7
  - Straight Chain Hydrocarbons in Kerosene, Gas Oils and Lubricating Oils 12
  - Normal Paraffins with 14 to 21 Carbon Atoms 14
  - Linear Paraffinic Hydrocarbons 18
  - Liquid Aliphatic Hydrocarbon Fermented by *Mycobacterium phlei* 21
  - Use of a Heavy Distillate Fraction 24
  - Use of Low Boiling and High Boiling Hydrocarbon Fractions 24
  - Normal Aliphatic Hydrocarbon Feed Used with Selected Microorganisms 26
- Other Liquid Feedstocks 29
  - Hydrocarbon-Derived Products Containing Aldehydes 29
  - Oxygenated Hydrocarbons 32
  - Growth on an Extract of Coal 34
  - Protein from Cyclohexane Oxidation Products 37
  - Mixture of Aliphatic Mono- and Dicarboxylic Acids 39

GASEOUS AND SOLID HYDROCARBON FEEDSTOCKS 42
- Gaseous Hydrocarbons 42
  - Butane 42
  - Propane and Butane in Natural Gas Fractions 45
  - Methane and Natural Gas 47
  - Liquefied Petroleum Gas 49
- Solid Hydrocarbons 55

GROWTH PROMOTING TECHNIQUES 59
- Use of Chemicals 59
  - Effect of Zinc on Yeast Growth and Protein 59
  - Addition of Alcohol and Fermentation Waste Liquor 61
  - Unsaturated Organic Acids 64
  - Nonionic Surface Active Agent 68
  - Use of Ozone 71
  - Use of Antibiotics 73
- Use of Dispersions 76
  - Effect of Finely Divided Adsorbant 76
  - Increasing Microbial Fermentation with Zeolites 79
  - Emulsification by Nonionic Surface Active Agents 81
  - Water-in-Oil Emulsions 83

Atomizing a Liquid Phase Feedstock 86
Other Techniques 91
Dialysis to Improve Cell Growth 91
Reducing the Vapor Pressure of Volatile Hydrocarbons 93
Carbohydrate Pretreatment Feedstock 96
Adaptation to Growth on Hydrocarbons 98

FERMENTATION TECHNIQUES 102
Specialized Apparatus 102
Control of Residence Time in the Reactor 102
A Two Step Process for Growth of Yeast and Bacteria 109
A Horizontal Cylindrical Chamber 113
Continuous Microbial Process 121
Specialized Techniques 127
Anaerobic Photosynthetic Fermentations 127
Maturation Treatment at Low Temperatures 129
Starvation Phase Effecting a Low Lipid Content 132
Controlling the Introduction of Additional Hydrocarbon Substrate 133
Reduction of Nucleic Acid Content of Cells 140

RECOVERY TECHNIQUES 144
Separation of Cells from Fermentation Media 144
Controlled Froth Flotation 144
Use of Microfine Air Bubbles 156
Addition of Immiscible Organic Liquid 159
Application of Surface Active Agents 161
Coagulation by Low pH and Heat 165
Phase Separation by Alcohol and Hydrocarbon 170
Phase Separation by a Dialkyl Ketone and Acetone or Alcohol 173
Phase Separation in a Gravitational Field 176
Removal of Residual Hydrocarbon 178
Application of a Water-Soluble Surfactant 178
Use of a Two-Component Combination of Surface Active Agents 182
Refermentation Techniques and Solvent Extraction 183
Solvent Extraction 190
Use of a Drum Dryer 193
Solvent Removal by Steam 196
Recovery of Proteins 199
Enzymatic Rupture of Cell Walls 199
Heating in an Aqueous Acidic Solution 202
Acid Treatment with Perchloric and Trichloroacetic Acids 206
Solvent Extraction of Lipids 209
Precipitating Extracted Protein at Acid pH 211

COMPANY INDEX 217
INVENTOR INDEX 218
U.S. PATENT NUMBER INDEX 220

Introduction

Microorganisms as a source of supply of man's requirements for protein is a subject of considerable study. The ability to achieve this goal has necessitated an adequate and inexpensive source of feedstock. A worldwide need for food has led to the evaluation and development of hydrocarbon feedstocks for the microbiological production of proteins.

Fermentation for the production of foods has been known for centuries. An understanding of the process has been realized only during the past century with the most intricate details of fermentation appearing within the last two to three decades. Presently, fermentations utilizing a wide variety of substrates as nutrients and a vast number of microorganisms, including bacteria, yeasts and molds, are producing antibiotics, steroids, vitamins, enzymes, amino acids, organic solvents, food chemicals, etc.

Many of these products are being produced because microbiologists have found ways of controlling the fermentation process. This control has been affected at times by man's choice of substrate and/or by man's choice of microorganism. The substrate may specifically determine the product of the reaction or it may simply be a choice to generate maximal growth. The microorganism may also determine the product of the reaction and may do so regardless of the nutrients selected for its medium. The goal may be a mass of cells or a specific product contained within the cells or available in the fermentation broth.

In some instances, production of a product by fermentation will have organic synthesis as a competitive procedure. However, when the desired product is a macromolecule, man's ingenuity cannot duplicate the speed, efficiency and competence of the microorganism. The synthesis of proteins, high molecular weight polymers composed of amino acids, is much too complex for the chemist but is an easy task for the microorganism when combined with the knowledge applied by the microbiologist.

The production of proteins from hydrocarbon feedstocks is already a reality in many parts of the world. Interest in this source of protein increases as the world population explosion continues. Hydrocarbons as a feedstock offer the advantage of being in large supply, readily transportable from source to a point where the end product, protein, is most needed, and available throughout the year without the risk associated with the growing and harvesting of food crops.

The fermentation process can be undertaken almost anywhere in the world. A food supply based on proteins derived from hydrocarbons through fermentation can be a reality in parts of the world where the raising of livestock or the growing of field crops is impossible or the output is inadequate. Furthermore, the rate of growth for microorganisms greatly exceeds the rate of growth of animals and food crops.

Fermentation has the further advantage of producing a large mass within a small land area when compared with the requirements of ranching or farming. However, a plant designed for the fermentation of hydrocarbons presents engineering problems for the management of the hydrocarbon as a feedstock and as a residue in the product. The final product must have acceptable quality. Toxicity due to residues does not permit use of the protein in man's diet or use of the dried cells as animal feed. Palatability is a consideration and growth conditions can lend to the accumulation of undesirable cellular components.

In treating the subject of proteins from hydrocarbons, the growth of cells on a variety of hydrocarbon feedstocks is presented. The solution of problems associated with these fermentations is described in terms of growth promoting procedures, specialized apparatus and techniques permitting the use of hydrocarbon feedstocks, separation of the cells from the fermentation media, removal of residual hydrocarbons from cells and recovery of protein from cells.

Large-scale commercial production of proteins from hydrocarbons will be a reality in the late 1970's. Several oil companies are involved in this undertaking. Nevertheless, research and development continue as these future manufacturers seek the best procedures for the production of proteins from hydrocarbons.

Liquid Hydrocarbon Feedstocks

PARAFFINIC FEEDSTOCKS

Straight Chain C_{10} Paraffinic Hydrocarbons

A. Champagnat and J. Filosa; U.S. Patent 3,193,390; July 6, 1965; assigned to The British Petroleum Company Limited, England have described an industrial process for the manufacture of food yeasts starting from petroleum fractions.

The process for the production of food yeast comprises maintaining a yeast in a nutrient medium in the presence of a paraffinic feedstock of petroleum origin and a free oxygen-containing gas, preferably air, under conditions favoring growth of the yeast. The feedstock has an average molecular weight corresponding to at least 10 carbon atoms per molecule.

While the feedstock should be rich in paraffinic hydrocarbons and preferably rich in normal paraffins, the feedstock may also contain minor amounts of other hydrocarbons, particularly naphthenes or aromatics; in general these hydrocarbons will not take part in the growth process and will be present in the product. If desired, an initial feedstock containing substantial amounts of paraffins together with other hydrocarbons may be treated for the separation of a paraffinic feedstock for use in the process.

Thus there may be used straight run fractions, for example, kerosenes; gas oils; middle distillate fractions, preferably boiling in the range of 150° to 450°C. Alternatively, there may be used treated fractions, for example, slack wax or other wax fractions, for example as are derived by the dewaxing of lubricating oil fractions; and paraffins derived by processes of separation by the use of molecular sieves.

In general the paraffinic feedstock will not require special treatment to adapt it for use in the process and in general there may be used low value fractions available as by-products of normal refinery operations.

Preferably the yeast which is grown on the paraffinic feedstock is of the family Cryptococcaceae and particularly of the subfamily Cryptococcoideae; however, if desired there may be used, for example, ascosporogeneous yeasts of the subfamily Saccharomycoideae. Preferred genera of the Cryptococcoideae subfamily are Torulopsis (also known as Torula) and Candida. Preferred strains of Candida are *Candida tropicalis* and, in particular, *Candida lipolytica* (also known as *Mycotorula lipolytica*).

When starting with a fresh stock of the yeast it will usually be necessary to adapt the yeast to assimilate carbon from hydrocarbons and to use an inoculum of the adapted yeast for the process of growth.

The growth of the yeast takes place at the expense of the petroleum fraction with the intermediate production of bodies having an acid function, principally fatty acids, in such manner that the pH of the aqueous mineral medium progressively diminishes. If one does not correct it the growth is fairly rapidly arrested and the concentration of the yeast in the medium, or cellular density, no longer increases.

It has been found that by the use of yeast of the strain *Candida lipolytica* in a process in which aeration is effected by "vortex aeration," a high growth rate is achieved whereby the generation time lies in the range of two to five hours and the cell concentration is increased by a factor of up to 12 g./l. in two days.

Example 1: Preparation of the Inoculum — One starts with a strain of *Candida lipolytica* previously grown on malt agar medium which must first of all be adapted to assimilate carbon from the hydrocarbons. A culture of *Candida lipolytica* is made in 250 ml. sterile Erlenmeyer flasks containing 50 ml. of mineral medium to which is added 1/10 ml. of n-hexadecane.

The mineral medium has the following composition:

Monopotassium phosphate	7 g.
Magnesium sulfate	0.2 g.
Sodium chloride	0.1 g.
Ammonium chloride	2.5 g.
Tap water (oligo-elements)	100 ml.
Yeast extract, Difco	1 mg.
Distilled water balance to 1,000 ml.	

This culture is incubated for three days with mechanical agitation at 30°C.

5 ml. of this culture is then removed and added to another flask containing the same quantities of the same mineral medium and hydrocarbon. The incubation is carried on for three days with agitation. The same operations are conducted several times, for example, four or five. Two 2-liter toxin-sterile flasks, each containing a liter of the above mineral medium to which has been added 2 ml. of n-hexadecane, are seeded with 20 ml. per liter of an inoculum prepared by the preceding culture. This culture is incubated for 36 hours at 30°C. with mechanical agitation. The 2 liters of culture are centrifuged at 30°C. and the yeasts collected constitute the inoculum.

Culture of Candida lipolytica on a Paraffinic Petroleum Product – The hydrocarbon feedstock utilized is the crude paraffinic product obtained by the dewaxing of a spindle oil obtained from a Kuwait crude petroleum. Its properties are:

Melting point (ASTM D127)	42.8°C.
Kinematic viscosity at 100°C.	3.02 cs.
Oil content (ASTM D127)	17.85%

1 mg./l. of yeast extract Difco is added to the same mineral nutrient medium as described above. The pH is adjusted to 5 at the beginning of the process. 290 ml. of the mineral medium containing the yeast extract are placed in a glass vessel of 1 liter equipped with an effective air injection device. The injection of air is started at the rate of 400 l./hr. and the temperature is maintained at 30°C., evaporation being compensated for by periodic addition of distilled water.

The yeast inoculum is added to the vessel and is maintained in suspension, whereupon the above paraffinic product is introduced by successive additions of 1 ml./hr. At the beginning the cellular density of the yeasts or concentration is 2 g./l. expressed in dry weight. By periodic removals of 25 ml., one measures the evolution of the cellular density of the culture. To this end, the yeast cells in the samples taken are filtered and washed until ammonium ions have disappeared, and dried, and then their nitrogen content is measured by the method of Kjeldahl. Measurements on pure and dried *Candida lipolytica* previously carried out indicate that its content of nitrogen is 7.25%.

From the nitrogen content of the sample taken one can deduce the cellular density of the culture at the moment of sampling. It has been found that the growth, as expressed by the increase of the cellular density, continues for about 9 hours, at the end of which one reaches a stationary phase. At this moment, the cellular density is 3.5 g./l. of the culture medium. The production of yeast per liter of medium and per hour is then

$$\frac{3.5 - 2}{9} = 0.166 \text{ g./l./hr.}$$

Example 2: Growth was carried out in exactly the same manner as Example 1. The culture of *Candida lipolytica* at the beginning of the fermentation had a cellular density of 1.60 g. of yeast per liter of medium by dry weight and a pH of 5. After 8 hours, approaching the stationary phase and with the pH having fallen to 3, the pH was adjusted to 5 by the addition of caustic soda lye.

The growth then commenced again and was continued for the following 5 hours until a cellular density of 6 g./l. was reached.

The production of yeast per liter of culture medium per hour was

$$\frac{6 - 1.6}{8+5} = 0.34 \text{ g./l./hr.}$$

Heavy Gas Oils

A process for the production of an edible yeast product with a concurrent dewaxing of a petroleum fraction containing initially from 3 to 45% by weight of n-paraffin has been described by *A. Champagnat and J.A. Filosa; U.S. Patent 3,560,341; February 2, 1971; assigned to The British Petroleum Company Limited, England*. The process produces a dewaxed gas oil product rendered more suitable for industrial use through a reduced proportion of n-paraffin and an edible yeast product which is cultivated at the expense of the n-paraffin initially present.

Petroleum fractions boiling below the gas oils, for example, heavy naphthas and kerosenes contain straight chain hydrocarbons which are potentially valuable for conversion to other products but, utilization of these hydrocarbons has been rendered difficult by the necessity of recovering these hydrocarbons from the petroleum fractions, in which they are contained, before they can be converted to other products.

It is also well known that food yeasts, suitable for consumption by cattle, or in some cases by human beings, are prepared using as starting materials carbohydrates.

This process comprises cultivating a strain of yeast adapted to grow on straight chain paraffinic hydrocarbons, in the presence of a petroleum fraction consisting in part of straight chain hydrocarbons and having a mean molecular weight corresponding to at least 10 carbon atoms per molecule.

When cultivating yeasts in the presence of the feedstocks under conditions favoring the growth of the yeasts at the expense of the straight chain hydrocarbons, the other hydrocarbons, for example isoparaffins, naphthenes and aromatics, are not metabolized or, at most, the proportion which is metabolized is very small. Furthermore, unlike conventional chemical process governed by the law of mass action, the rate of removal of straight chain hydrocarbons is not substantially reduced as the proportion of these hydrocarbons in the overall mixture of hydrocarbons decreases (except, of course, in the very final stages of removal).

Thus, the percentage conversion of straight chain hydrocarbons which is achieved can be maintained at a value approaching 100% without necessitating a very disproportionate expenditure of content time to achieve small improvements. Furthermore, in a continuous process, this high percentage conversion can be achieved without resorting to the use of a long reaction path.

Example 1: The feedstock was a heavy gas oil having the following characteristics.

		Methods
Density at 15° C./4° C	0.866	NFT60101
Distillation, initial point, ° C	204	NFM07009
Percent at 250° C	2	
Percent at 270° C	6	
Percent at 357° C	82	
Final point, ° C	370	
Pour point, ° C	+9	NFT60105
Content of normal paraffins by wt., percent	18	

The yeast employed was a strain of *Candida lipolytica* previously cultivated on a malt sugar medium and adapted to withdraw its carbon from hydrocarbons. This adaptation was carried out by preparing an inoculum by the method now described.

Preparation of the Inoculum – A culture of *Candida lipolytica* is made in sterile 250 ml. conical flasks. These are open to the air and contain 50 ml. of a mineral medium together with 5 to 10 ml. of the heavy gas oil described above. The mineral medium has the following composition:

Monopotassium phosphate	7 g.
Magnesium sulfate	0.2 g.
Sodium chloride	0.1 g.
Ammonium chloride	2.5 g.
Tap water (trace elements)	100 ml.
Difco yeast extract	1 ml.
Distilled water up to 1,000 ml.	

This culture is incubated for three days at 30°C. with mechanical agitation.

5 ml. of this culture are then taken and added to a flask containing the same quantities of the same mineral medium and hydrocarbons. The incubation is continued for three days with agitation. These same operations are repeated four or five times.

Two sterile 2-liter toxin flasks, each containing 1 liter of the above mineral medium to which 10 ml. of heavy gas oil have been added, are seeded by means of 20 ml./l. of an inoculum coming from the preceding culture. This culture is incubated for 36 hours at 30°C. with mechanical agitation. The 2 l. of culture are centrifuged at 30°C. and the yeasts recovered in the centrifuge cups constitute the inoculum necessary for one experiment. Subsequent stages were carried out in the following manner.

Culture on Heavy Gas Oil – There is introduced into a fermenter, equipped with an aeration system of a type which forms a vortex in the mixture, 1 l. of the same mineral nutritive medium as described above (containing 1 ml./l. of yeast extract). The pH is regulated to 4.5 at the beginning of the experiment and is maintained at this figure during the culture by the addition of monopotassium phosphate.

The air injection is started up at the rate of 50 l./hr./l. of the medium and the temperature is maintained at 30°C., and periodically distilled water is added to compensate for evaporation losses.

The fermenter is then seeded with 800 mg. of the inoculum whose preparation is described above, and the gas oil is introduced in four stages in quantities in an exponentially increasing series, so as to reach in all 25 ml./l. of medium.

The incubation is stopped when the cellular density (concentration of cells) remains constant, that is to say when the stationary phase has been reached. The cellular density is measured by taking periodical 25 ml. samples of the culture, the cells of which are filtered, washed until the NH_4 ions disappear, and dried. Their content of nitrogen is estimated by the Kjeldahl method. Measurements carried out previously on pure dry *Candida lipolytica* gave a nitrogen content of 7.25%. From the nitrogen content of the test sample taken it is possible to deduce the cellular density of the culture at the moment the sample was taken.

Recovery of the Yeasts – To the emulsion formed there is added 0.025% by weight, reckoned on the aqueous phase of the cationic surfactant, ammonium stearyl-trimethyl chloride and the pH is adjusted to 8 by the addition of caustic soda. The discontinuous centrifuging of this mixture gives:

(a) a pasty phase of yeast cells impregnated with hydrocarbon and aqueous medium;

(b) a limpid phase of aqueous mineral medium;

(c) a wet hydrocarbon phase.

The pasty phase containing the yeast cells is then washed with water containing 0.25 part per thousand of the same surfactant. Two washings and two centrifugings are sufficient to obtain yeast which is free from hydrocarbons. Finally the yeast is washed with pure water at

60°C. to eliminate the surfactant left in the yeast and it is then centrifuged. The product thus obtained is dried at 80° to 90°C. in a rapid current of air.

Recovery of the Dewaxed Heavy Gas Oil — To the wet hydrocarbon phase from the first centrifuging is added the wet hydrocarbon phases from the two subsequent centrifugings and the mixture is allowed to settle at 30°C., until the water which it contains separates out. (More satisfactorily this step is carried out by centrifuging.)

The table below gives the numerical results obtained in this experiment.

Initial cellular density	0.800 g./l.
Final cellular density	4.76 g./l.
Growth time	11 hr.

Production of yeast per liter of culture medium and per hour:

$$\frac{4.76-0.80}{11} = 0.360 \text{ g./l./hr.}$$

Yield of yeast in relation to the gas oil:

$$\frac{4.76-0.80}{25 \times 0.866} = 18\% \text{ by weight}$$

Yield of dewaxed gas oil: 82% by weight
Pour point dewaxed gas oil: -40°C.

	Feedstock Reference						
	A	B	C	D	E	F	G
Nature of gas-oil	Straight run Zarzaitine	Irak	Kuwait	Irak	Heavy Zarzaitine	Feedstock after ferrofining	Feedstock after desulphurisation by hydrofining
Characteristics of gas-oil before treatment, distillation:							
Initial B.P., °C	305	224	198	223	178	305	223
50%, °C	327	348	361	366	380	327	366
Final B.P., °C	351	390	400	400	400	351	400
Pour point, °C	+8	+11	+17	+22	+26	+6	+21
Content of normal paraffins percent by weight	13.3	13.2	8.8	14	11.6	13.3	12
Characteristics of gas-oil after treatment:							
Pour point, °C	−25	−24	−20	−16	+3	−25	−9.3
Percentage or normal paraffins but metabolised (by weight)	0.18	0.5	0.87	1.4	3.4	0.1	1.5
Yield of yeast, based on feedstock (by weight)	12.5	13.2	8.1	7.8	8	16	11.5

Straight Chain Hydrocarbons in Kerosene, Gas Oils and Lubricating Oils

Certain petroleum fractions, particularly gas oils, contain straight chain hydrocarbons, mainly paraffins which are waxes and which have an adverse effect upon the pour point of the fraction; that is to say, when these hydrocarbons are removed wholly or in part, the pour point of the fraction is lowered. Usually the wax is removed by precipitation by means of solvents, the wax originally present in the fraction being recovered as such, that is, without conversion to more valuable products.

The petroleum fractions boiling below the gas oils, for example, heavy naphthas and kerosenes, also contain straight chain hydrocarbons which are potentially valuable for conversion to other products. Hitherto, in general, utilization of these hydrocarbons has been rendered difficult by the necessity of recovering these hydrocarbons from the petroleum fractions in which they are contained before they can be converted to other products.

The process of *A. Champagnat and C. de Mayo; U.S. Patent 3,257,289; June 21, 1966; assigned to The British Petroleum Company Limited, England* provides for the production of yeasts from hydrocarbons, for the removal wholly or in part of straight chain hydrocarbons from petroleum fraction with the recovery of a valuable product based on the original straight chain hydrocarbon content of the petroleum fractions and for the dewaxing of a petroleum gas oil fraction with the production of edible yeast.

There is provided a process which comprises the steps of continuously cultivating, in a fermenter, a strain of yeast which is adapted to grow on straight chain paraffinic hydrocarbons, in the presence of a petroleum fraction consisting in part of straight chain hydrocarbons and having a mean molecular weight corresponding to at least 10 carbon atoms per molecule.

The process is of particular value for the treatment of petroleum gas oil fractions which contain straight chain hydrocarbons in the form of waxes, since by this process, a gas oil of improved pour point is obtained while the waxes are converted to yeast.

Suitable feedstocks to the process include kerosene, gas oils and lubricating oils; these feedstocks may be unrefined or may have undergone some refinery treatment, but must contain a proportion of straight chain hydrocarbons. Suitably the petroleum fraction will contain 3 to 45% by weight of straight chain hydrocarbons. The yeast grown on the feedstock is of the family Cryptococcaceae and particularly of the subfamily Cryptococcoideae.

The growth of the yeast used is favored by the addition to the culture medium of a very small proportion of "bio" factors, for example extract of yeast (an industrial product rich in vitamins of group B obtained by the hydrolysis of a yeast) or more generally of vitamins of group B and/or biotin. This quantity is preferably of the order of 25 parts per million with reference to the aqueous fermentation medium. It can be higher or lower according to the conditions chosen for the growth.

At the end of the growth one obtains an emulsion which contains yeast and nonmetabolized hydrocarbons in a continuous aqueous phase.

The major part of the continuous aqueous phase is first separated; preferably this is carried out by centrifuging. The separated aqueous phase will usually contain a greater concentration of non-nutritive ions that can be tolerated in the recycle stream and when this is so, only a proportion of the recovered aqueous phase can be recycled. Thus it will usually be possible to separate ca. 96% by weight of the aqueous phase which is present in the product, of which on the same percentage basis, ca. 20% by weight, will be discarded. The recycle stream is supplied with makeup quantities of the necessary nutrients and is returned to the fermenter; if desired the makeup materials may be fed to the fermenter as a separate stream.

By centrifuging the product from the fermenter three fractions may be recovered. These are in order of increasing density:

(a) an oil phase containing yeast cells;
(b) an aqueous phase containing traces of oil and yeast; and
(c) a yeast "cream" consisting of yeast, having a quantity of oil fixed onto the cells, together with aqueous phase.

After recovery of fraction (b), the remaining fractions (a) and (c) may be reblended and mixed with an aqueous solution of a surfactant. The purpose of this treatment is to separate the oil from the yeast cells; the oil being apparently held to the cells by adsorption.

It may be advantageous to employ an edible surfactant, for example, a saccharose ester, which makes it possible to reduce the subsequent washing required to remove from the yeast a surfactant which is not edible.

The emulsion so formed is broken down by centrifuging to obtain three fractions:

(d) an oil phase;
(e) an aqueous phase containing surfactant, which phase is recycled for the treatment of fractions (a) and (c);
(f) a yeast cream, consisting of yeast still contaminated by oil together with an aqueous surfactant phase.

In order to reduce as far as possible the consumption of surfactant product, the aqueous washing solution containing it is recycled.

Fraction (f) may be further treated by alternate washing with surfactant and centrifuging until the oil content of the yeast has reached a desired low value. The yeast cream now consisting of yeast and aqueous surfactant may now be washed with water and again centrifuged.

Normal Paraffins with 14 to 21 Carbon Atoms

Y. Nakao, I. Banno and M. Kuno; U.S. Patent 3,411,989; Nov. 19, 1968; assigned to Takeda Chemical Industries, Ltd., Japan have described methods for preparing yeast cells containing an enhanced amount of ribonucleic acid, which comprises inoculating a mutant of genus Candida, which mutant is resistant to thiazine dye and/or oxazine dye and/or acridine dye, onto a culture medium wherein the carbon source consists mainly of hydrocarbons containing not less than 10% (volume/volume) of normal paraffins having a carbon atom number from 14 to 21, the culture medium also containing other nutrients necessary for the growth of the mutant, and incubating the culture medium.

Yeasts belonging to genus Candida can especially digest normal paraffins of carbon atoms numbers within a range from 14 to 21 which are readily available at a relatively low cost as carbon source among many kinds of hydrocarbons. Further, when mutants which are induced from strains of genus Candida and which are resistant to certain dyestuffs including thiazine dye, oxazine dye and acridine dye, are incubated in a culture medium wherein the carbon source consists mainly of hydrocarbons containing the said normal paraffins, cells containing a considerably enhanced amount of ribonucleic acid can be obtained.

In sharp contrast with the above, cells containing only a limited amount of ribonucleic acid are obtained when wild strains of these yeast are incubated in the culture medium. Moreover, the mutants resistant to the dyestuffs are obtainable as mutants induced artificially or spontaneously from those naturally existing strains whose growth is inhibited by the presence of the dyestuffs.

The principal object is a process for preparing yeast cells containing a large amount of ribonucleic acid with the use of cheap and easily available hydrocarbons, on an industrial scale.

The object is realized by incubating mutant of genus Candida which is resistant to a dyestuff, i.e., thiazine dye, oxazine dye or acridine dye, in a culture medium wherein the carbon source consists mainly of hydrocarbons containing normal paraffins of any carbon atom number within the range from 14 to 21, in the presence of other nutrients required for its growth.

The amount of the normal paraffins of carbon atom numbers within the range of 14 to 21 contained in the hydrocarbons is not less than 10 volume percent of the latter, from the viewpoint of both the growth of the mutant and amount of ribonucleic acid accumulated in their cells. The normal paraffins themselves are advantageously employed as 100% of the hydrocarbon source.

The normal paraffins may be either those which consist of only one kind of normal paraffin of a carbon atom number within the range of 14 to 21 or those which consist of two or more kinds of normal paraffins, each having a carbon atom number within that range. The hydrocarbon source may contain other hydrocarbons (e.g., branched paraffins, olefins, cyclic paraffins, aromatic hydrocarbons, normal paraffins of 1 to 13 or of not less than 22 carbon atoms, etc.) than the normal paraffins of carbon atom numbers within the range of 14 to 21, as long as the normal paraffins of carbon atom numbers within the range of 14 to 21 are included as a whole in an amount of not less than 10% (vol./vol.).

As such hydrocarbons, there are preferably employed pure normal hexadecane, heavy paraffins which contain about 90% (vol./vol.) of normal paraffins of carbon atom numbers within the range of 14 to

21 and have a boiling range of from about 262° to 349°C., heavy gas oil which contains about 17% of normal paraffins of carbon atom numbers within the range of 15 to 20 and has a boiling range from about 272° to 386°C., and gas oil which contains about 14% of normal paraffins of carbon atom numbers, within the range of 14 to 17 and has a boiling range from about 180° to 350°C.

From the viewpoints of both the growth of the mutant and the amount of ribonucleic acid accumulated in its cells, the hydrocarbons are generally used in such an amount as to make the concentration in the culture medium of the normal paraffins of carbon atom numbers with the range of 14 to 21 as a whole about 3 to 20% (vol./vol.).

As these hydrocarbons are scarcely soluble in water, the addition to an aqueous culture medium is practically carried out under stirring or shaking to prepare a suspension containing very fine particles. If desired, a suspending agent, e.g., a surfactant of the type of polyoxyethylene sorbitan monostearate (commercially available as Tween 60) may be employed. These hydrocarbons are by themselves sufficient carbon sources but, if desired, commonly employable carbon sources such as carbohydrate (e.g., glucose) may be used together with the hydrocarbons.

Example: Candida tropicalis Berkhout is inoculated in 10 ml. of the culture medium prepared by supplementing the composition according to the following table with 400μg./ml. of tetramethyldiaminoxanthylium chloride, and the medium is incubated under shaking at 28°C. for four days.

Sucrose	50 g./l.
Sodium glutamate	3 g./l.
Ammonium sulfate	25 g./l.
KH_2PO_4	1 g./l.
$MgSO_4 \cdot 7H_2O$	0.5 g./l.
NaCl	0.1 g./l.
$CaCl_2$	0.1 g./l.
Biotin	2 μg./l.
Calcium pantothenate	400 μg./l.
Inositol	2,000 μg./l.
Nicotinic acid	400 μg./l.
p-Aminobenzoic acid	200 μg./l.
Pyridoxine hydrochloride	400 μg./l.
Riboflavin	200 μg./l.
Thiamine hydrochloride	400 μg./l.
Water up to 1,000 parts by volume (pH 5.8)	

From the resultant culture broth spontaneously occurring tetramethyldiaminoxanthylium chloride-resistant mutants are isolated.

Among those mutants isolated, a strain (designated as *Candida tropicalis* Berkhout-Ca3-Py) (ATCC No. 20006) is inoculated in 50 parts by volume of the culture medium according to the following table, and the medium is incubated under shaking at 28°C. for 24 hours.

	Parts by Weight
Heavy paraffins[1]	100
NH_4NO_3	10
KH_2PO_4	20
K_2HPO_4	10
$MgSO_4 \cdot 7H_2O$	1
$MnSO_4 \cdot 4H_2O$	0.01
$ZnSO_4 \cdot 7H_2O$	0.01
$FeSO_4 \cdot 7H_2O$	0.01
Cornsteep liquor	10
Water up to 1,000 parts by volume. (pH 6.0)	

[1] Having the following properties: (a) Components—normal paraffins of carbon atom numbers within the range of 14 to 21, 90.7%; normal paraffins of carbon atom numbers within the range of 11 to 13, 1.9%; and normal paraffins of carbon atom numbers within the range of 22 to 23, 8.5%. (b) Boiling range— 262° to 349°C.

The resultant culture broth is inoculated in 1,000 parts by volume of the culture medium of the same composition as that in the above table, and incubated under shaking at 28°C. for 20 hours. The resultant culture broth is subjected to centrifugation to give 24 parts by weight (as dry matter) of cells containing 12% of ribonucleic acid relative to the whole weight. The cells are suspended in 250 parts by volume of 0.5 M sodium chloride solution adjusted to pH 8.0, and the suspension is heated at 90°C. for 30 minutes to extract ribonucleic acid. After the cells are eliminated by filtration, 100 parts by volume of ethanol is added to the resultant filtrate to yield 2.2 parts by weight of ribonucleic acid.

When a wild strain of *Candida tropicalis* Berkhout is incubated instead of the said mutant, *Candida tropicalis* Berkhout Ca3-Py (ATCC No. 20006) under the same conditions as above, 20 parts by weight (as dry matter) of cells containing only 4.0% of ribonucleic acid relative to the whole weight are obtained.

Linear Paraffinic Hydrocarbons

In the cultivation of microorganisms, as described by *R. Avrillon, S. Franckowlak, C. Gatellier and G. Glikmans; U.S. Patent 3,698,998; October 17, 1972; assigned to Institut Francais du Petrole des Carburants et Lubrifiants, France,* a liquid hydrocarbon feedstock is employed containing at least 70% by weight of linear paraffinic hydrocarbons. The growing step is conducted in the presence of a gas containing molecular oxygen and with a nutrient aqueous medium containing:

(a) assimilable nitrogen and phosphorus;

(b) oligoelements such as trace amounts of copper, cobalt, iron, molybdenum and manganese compounds;

(c) growth factors such as yeast, etc; and

(d) essential mineral elements such as potassium, ammonium and magnesium ions.

The microorganisms are then recovered from at least the major part of unconverted hydrocarbons, preferably by decantation, and then subjected to a maturation step in an aqueous medium similar to the nutritive aqueous medium, but deprived of at least one of the essential mineral elements.

The first step consists of dewaxing a starting hydrocarbon cut so as to obtain an effluent containing at least 70% and preferably at least 85% by weight of normal (linear) paraffins.

The effluent will, for instance, contain about from 75 to 95% by weight of normal paraffins. However the normal paraffin content will generally not exceed 99%. A content of more than 95% and specially more than 99%, results in a significant cost increase of the first step which will have a substantial effect on the overall cost of the process.

This starting hydrocarbon cut is generally a cut of the kerosene type, and oily cut or preferably a gas oil cut. The dewaxing of this starting cut will be carried out preferably according to the process described in the French Patent 1,463,784. This process consists of contacting a hydrocarbon cut with urea in suspension or dissolved in a mixture of nitrile, monocarboxylic acid and water. There is formed a solid complex of urea with normal paraffins, that complex being separated from the residual hydrocarbon and from the major part of the water-nitrile mixture in one or more separating units, in most cases centrifugation devices, preferably of the cyclone type, those complexes being afterwards destroyed by heat so as to recover the desired normal paraffins.

It is critical to obtain the abovementioned n-paraffins contents at the end of the first step since a feedstock containing less than 70% of n-paraffins cannot be treated beneficially; in particular the maturation step does not produce the desired result as shown hereinafter more in detail.

During the maturation step, the microorganisms no longer multiply. They oxidize and convert at first the hydrocarbon(s) still adsorbed on the walls of their cells and thereafter convert the lipids and fatty substances both external (adsorbed) or internal to their cells.

The maturation step therefore results in a decrease of the lipid and fatty substance content both of the extra- or intracellular type (generally up to a substantially complete consumption of the lipids) and also in an increase of the protide content of said microorganisms.

Example: There are contacted at 20°C. a gas oil containing 11% by weight of linear paraffinic hydrocarbons with a mixture of urea, acetonitrile and water, according to the process described in French Patent 1,463,784.

This results in the formation of a preferential complex of linear paraffins with urea; the complex is separated from the remaining hydrocarbons in a series of hydrocyclones.

The complex is then decomposed by heating at 70°C. There are thus recovered a hydrocarbon phase containing 89% by weight of linear paraffins (the complement to 100% consisting essentially of isoparaffins) and, naturally, urea.

This hydrocarbon phase is then contacted, in a fermentation vessel at 30°C., with an inoculum of *Candida lipolytica,* preliminarily cultivated on hydrocarbons, in the presence of an aqueous nutritive medium of the following composition, and in the presence of air in a divided form. The composition of the nutritive aqueous medium is as follows.

	Grams
Ammonium acid phosphate	2.9
Magnesium sulfate	0.9
Potassium chloride	1.32
Yeast extract (growth factor)	0.03
Spring water (containing oligoelements)	300
Distilled water	700

The pH value of the medium is maintained at 4 by ammonia injection. The air flow is such as to form an emulsion of the three phases (hydrocarbon liquid phase, aqueous phase and gaseous phase) providing a good contact between the yeasts and these three phases.

The fermentation vessel is continuously fed with a hydrocarbon substrate containing 89% of linear paraffins, with aqueous nutritive medium and with air while continuously withdrawing a part of the culture medium.

The effluent is conveyed to a decanter where there quickly appear two phases: (1) a mixed phase containing water and hydrocarbons; (2) a yeast cream at the lower part of the decanter.

The yeast cream is conveyed to a maturation vessel (identical to the fermentation vessel) where it is contacted, in the presence of air, with a non-nutritive aqueous medium composed of:

	Grams
Ammonium acid phosphate	2.9
Potassium chloride	1.32
Spring water (oligoelements)	300
Distilled water	700

The temperature and pH conditions are identical to those prevailing in the fermentation vessel. The respective proportions are of 100 volumes of yeast cream per 16 volumes of aqueous medium.

By microscopic examination of the yeast cells, there is observed, at first, a consumption of the hydrocarbons previously adsorbed during the fermentation step by the cell walls, and subsequently a change in appearance of the cell cytoplasm.

After 12 hours of maturation, yeasts are recovered by centrifugation. They are thereafter washed three times with ordinary water and dried, they do not become rancid in air under usual storage conditions.

The following table gives the composition by weight of the cells, expressed as a function of time.

Time in hours	0	2	12
Nitrogen	6.1	6	8.5
External fatty substances plus adsorbed hydrocarbons	0.5	0	0
Internal lipides	10.5	10.4	1
Total of fatty substances	11.0	10.4	1

Liquid Aliphatic Hydrocarbon Fermented by *Mycobacterium phlei*

U.A. Phillips; U.S. Patent 3,627,637; December 14, 1971; assigned to Commercial Solvents Corporation has proposed the production of organic nitrogenous materials by cultivating a protein-producing and amino acid-producing strain of the microorganism *Mycobacterium phlei* in an aqueous, nutrient, fermentation medium containing as an energy source a normally liquid aliphatic hydrocarbon and an assimilable nitrogen source.

When the microorganism has reached a desirable amount of growth, the protein-containing cells of the microorganism are separated from the fermentation medium, and are then useful as a source of protein, particularly for the formulation of animal feed compositions. The separated protein-containing cells may also be dried and then used as a source of protein. L-glutamic acid produced by the microorganism is present in the fermentation medium and may be recovered. The new protein and amino acid-producing strain of *Mycobacterium phlei* is identified by the numerical designation NRRL B-3423.

It is surprising to obtain the advantageous results when using *Mycobacterium phlei* NRRL B-3423 with aliphatic hydrocarbons since it was not previously reported that *Mycobacterium phlei* could utilize a liquid aliphatic hydrocarbon as an energy source.

The liquid aliphatic hydrocarbon can be any normally liquid aliphatic hydrocarbon or mixture of hydrocarbons preferably having from about 6 to 20 carbon atoms in the molecule, and more preferably 10 to 16, and mixtures. Preferably the mixture consists principally of saturated, straight chain hydrocarbons. A satisfactory mixture can be obtained by the distillation of petroleum hydrocarbons. The fraction having an initial boiling point above about 95°F. and an end point below about 800°F. is useful as the hydrocarbon source.

The liquid aliphatic hydrocarbon may be one obtained by distillation of a crude oil or it may also be a fraction recovered from a thermal or catalytic cracking operation. Gasoline, naphtha, burning oils, kerosene, distillate fuels, gas oils and diesel fuels are suitable hydrocarbon fractions.

It is not necessary that the hydrocarbon fraction be free from aromatic, branched chain, and unsaturated aliphatic hydrocarbons; however, *Mycobacterium phlei* NRRL B-3423 does not generally provide satisfactory growth from aromatic, branched chain, and unsaturated aliphatic hydrocarbons. Although the preference of the microorganism is for normal alkanes, the presence of other types of hydrocarbons does not generally inhibit the growth of the microorganism nor the production of L-glutamic acid provided the normal alkanes are present in an amount sufficient to provide the energy necessary for satisfactory growth.

Kerosene is a specifically preferred liquid aliphatic hydrocarbon mixture. Kerosene is the hydrocarbon fraction usually having an initial boiling point of about 370°F. and an end point of about 620°F. with a gravity generally of about 35 to 50 API and is composed almost entirely of saturated hydrocarbons of from about 10 to 16 carbon atoms. The preference of the organism for this hydrocarbon mixture is shown by the manometric data below.

Oxidation of Hydrocarbons by

***Mycobacterium phlei* NRRL B-3423**

Substrate	Formula	Microliters of O_2 Consumed in 2.5 Hr.
Hexane	C_6H_{14}	281.1
Naphtha	C_8 to C_{10}	341.1
Decane	$C_{10}H_{22}$	497.7
Undecane	$C_{11}H_{24}$	468.6
Dodecane	$C_{12}H_{26}$	552.2
Tridecane	$C_{13}H_{28}$	641.8
Tetradecane	$C_{14}H_{30}$	684.1
Hexadecane	$C_{16}H_{34}$	720.4
Kerosene	C_{10} to C_{16}	584.4
Paraffin oil	C_{18} & above	434.8
Toluene	$C_6H_5CH_3$	66.2
Xylene (o, m, & p)	$C_6H_4(CH_3)_2$	91.5
Phenol	C_6H_5OH	34.5
Benzene	C_6H_6	31.5
Endogenous*		66.2

* Normal respiration of *Mycobacterium phlei* in absence of substrate

The kerosene or other liquid aliphatic hydrocarbon is generally employed in the fermentation medium in amounts of from about 0.1 to 10% by volume, but preferably from about 1 to 5% by volume.

It is preferred to employ along with the kerosene or other liquid aliphatic hydrocarbon a small amount of a dispersing agent to insure dispersion of the hydrocarbon throughout the aqueous fermentation medium. Suitable dispersants include cationic surface active reaction products of ethylene oxide and aliphatic amines having 10 to 30 carbon atoms per molecule, nonionic polyoxyethylene derivatives of hexitol anhydride, nonionic surface active alkyl aryl polyether alcohols having the general formula

$$R{-}(O{-}CH_2{-}CH_2)_n{-}OH$$

where R is an alkyl alcohol and n is an integer of about 5 to 100 or more. The amount of dispersant, if utilized, will fall in the range of about 0.00025 to 0.05% by weight, preferably 0.001 to 0.01% by weight.

Example: The inoculum was prepared by growing *Mycobacterium phlei* NRRL B-3423 for 18 hours at 30°C. on a medium having the following compositions:

Pancreatic digest of casein	17.0 g.
Papaic digest of soyameal	3.0 g.
Sodium chloride	5.0 g.
Dipotassium hydrogen phosphate	2.5 g.
Dextrose	2.5 g.
Water	1,000 ml.

The medium was agitated on a rotary shaker at 385 rpm throughout the 18 hour growth period.

A fermentation medium having the following composition was separately prepared:

Magnesium sulfate ($MgSO_4 \cdot 7H_2O$)	6.0 g.
Dipotassium phosphate	24.0 g.
Ammonium sulfate	48.0 g.
Manganese sulfate ($MnSO_4 \cdot H_2O$)	4 ppm
Ferrous sulfate ($FeSO_4 \cdot 7H_2O$)	4 ppm
Yeast extract	6.0 g.
Dispersing agent	0.090 g.
Water	5,7000 ml.
Kerosene	285 ml.

Approximately 3 liters of the above fermentation medium were placed in a laboratory fermentor and the contents were sterilized for 15 minutes at 115°C. and 10 psig pressure.

A 100 ml. portion of the inoculum was added to the fermentation medium in the laboratory fermentor and a fermentation was conducted at 30°C. for about three days. During the fermentation period the fermentor was agitated by a rotary stirrer at 1,750 rpm and the pH was adjusted at 24, 48, and 72 hours using normal ammonium hydroxide. Sterilized air was passed through the fermentation medium at the rate of 1.7 volumes per volume of medium per minute at a back pressure on the fermentor of 2 psig. This rate of airflow is equal to 5.1 liters per minute at 30°C. and 2 psig. The results of the fermentation are given below.

			Assay, g./l.	
Time, (hr.)	Adjusted pH	% Light Transmission	L-glutamic Acid	Proteinaceous Material, dry
24	7.3	87.0	0	–
48	7.7	72.5	2.5	–
72	8.8	49.0	5.5	16.4

Use of a Heavy Distillate Fraction

In the procedures of *B.M. Laine; U.S. Patent 3,268,414; August 23, 1966; assigned to The British Petroleum Company Limited, England*, a petroleum fraction is distilled to obtain at least two distillate fractions comprising a heavy distillate fraction and a light distillate fraction, thereafter treating the heavy distillate fraction with a microorganism whereby the microorganism grows using, as feedstock, straight chain hydrocarbons contained in the heavy distillate fraction, and thereafter separating the treated heavy distillate fraction. Microorganisms which are cultivated may be yeast, molds or bacteria.

Use of Low Boiling and High Boiling Hydrocarbon Fractions

I. Takeda and T. Iguchi; U.S. Patent 3,616,214; October 26, 1971; assigned to Asahi Kasei Kogyo K.K., Japan describe a method of production of L-glutamic acid, yeast and other useful substances using culture medium containing hydrocarbons such as petroleum fractions, paraffin oil and normal paraffin as the source of hydrocarbon and by adding higher boiling fractions of hydrocarbons in the beginning and then hydrocarbons containing lower boiling fractions.

In general, concerning the growing of microorganisms on hydrocarbons as the source of carbon, it has been reported that C_{10} to C_{13} n-paraffins or petroleum fractions containing them had little growing ability, and that good results could not be obtained in formation of L-glutamic acid, inosine or yeast.

It was found possible to form these useful substances in good yield by culturing microorganisms or yeast on n-paraffins or petroleum fractions containing them as the source of carbon, and then by adding n-paraffins of lower carbon numbers or petroleum fractions containing them.

A culture was continued thus: *Corynebacterium petrophilium* (ATCC No. 19,080), was inoculated into the culture medium consisting of n-hexadecane as the main source of carbon, as shown in Table 1, cultured about 15 hours under aerobic conditions, then added 4.5 g./dl. of C_8 to C_{13} n-paraffin and further 50 IU/ml. of penicillin G. The formation of L-glutamic acid was 2.5 g./dl. after 72 hours from the beginning.

TABLE 1

n-hexadecane	0.4 g.
Ammonium sulfate	0.3 g.
Urea	0.8 g.
Potassium dihydrogen phosphate	0.1 g.
Magnesium sulfate	0.05 g.
Ferrous sulfate	0.004 g.
Yeast extract	0.2 g.
Water (tap water)	100 ml.
pH	7.0

Also using the culture medium, as shown in Table 2, *Candida petrophilum* (ATCC No. 20,226), which was described in Japanese Patent '65-20666, was inoculated, cultured for 15 hours, and added 1.6 g./dl. of C_{10} to C_{13} n-paraffin, then cultured for 20 hours. The amount of dried microorganisms in the culture medium was 2 g./dl.

TABLE 2

n-hexadecane	0.4 g.
Ammonium sulfate	0.1 g.
Urea	0.2 g.
Potassium dihydrogen phosphate	0.1 g.
Magnesium sulfate	0.05 g.
Ferrous sulfate	0.004 g.
Yeast extract	0.1 g.
Water (tap water)	100 ml.
pH	5.0

Example 1: The culture medium shown in Table 1 was poured into 50 ml./500 ml. shaking flask, and cultured by shaking after *Corynebacterium hydrocarboclustus* M 104 (ATCC No. 21,402) was inoculated. After 15 hours, 4.6 g. of C_{10} to C_{13} n-paraffin, and then 30 IU/ml. of penicillin G were added. The amount of L-glutamic acid formed in the culture medium after 70 hours from the beginning was 1.5 g./dl.

On the other hand, only 0.5 g./dl. of L-glutamic acid formed when the fermentation was continued similarly, except the addition of 4.4 g. of C_{10} to C_{13} n-paraffin with 0.4 g./dl. of n-hexadecane in the beginning.

Example 2: 5 l. culture of medium shown in Table 2 was poured into a 10 l. jar-fermentor. *Candida tropicalis* was then inoculated, and cultured at 30°C., 500 rpm, at the aerating degree of 5 l./min. After 10 hours from the beginning, C_9 to C_{16} n-paraffin was added. Thus yeast was produced. Yeast was obtained by centrifugal separation and drying the culture solution. About 1 kg. of yeast was obtained from 1 kg. of hydrocarbon added.

Normal Aliphatic Hydrocarbon Feed Used with Selected Microorganisms

The culture of microorganisms on petroleum substrates to produce esters and chemicals as a major product and microbe cells as a by-product in very small amounts involves the use of less expensive carbon-containing feed materials, e.g., hydrocarbons rather than carbohydrates; but has not attained wide acceptance due to the difficulty of securing microbe cells having a high protein content coupled with a sufficiently high essential amino acid index. Other problems frequently connected with biosyntheses using hydrocarbon feed stocks are low cell growth rates (extremely long residence times) and inability of the microbe cells to effectively utilize hydrocarbon feeds for growth and reproduction.

J.D. Douros, Jr.; U.S. Patent 3,308,035; March 7, 1967; assigned to Esso Research and Engineering Company has obtained a marked improvement in protein biosynthesis by securing productive growth of microorganisms, in good yield at attractive growth rates, while using comparatively inexpensive C_1 to C_{30} aliphatic hydrocarbon feeds, e.g., C_1 to C_{30} n-paraffins, C_6 to C_{30} olefins, etc., for microbial growth. The microorganisms can be isolated readily from the biosynthesis bath in which they are grown.

Eight microorganisms useful in the process are tabulated below along with their corresponding ATCC registration number.

Microorganism Name	ATCC Number
Pseudomonas ligustri	15522
Pseudomonas pseudomallei	15523
Pseudomonas orvilla	15524
Alcaligenes sp.	15525
Cellumonas galba	15526
Brevibacterium insectiphilium	15528
Corynebacterium sp.	15529
Corynebacterium pourometabolum	15530

Each of these eight microorganisms has a valuable composite of properties, viz., a high protein content in excess of 50%, an essential amino acid index in excess of 45 and an excellent amino acid profile. Moreover, the microorganisms are nontoxic and thus can be used in animal feed supplements.

The protein can be extracted from these microorganisms by conventional extraction procedures and the protein extract then used as a glue or adhesive. A suitable protein extraction procedure involves sequential lysing, e.g., with acetone or other suitable organic lysing solvent, basic or acid extraction and isoelectric precipitation.

Intracellular and/or extracellular amino acids can be isolated from the microorganism cells and/or fermentation media. The process can be visualized as a combined process for biosynthesis of cells and chemicals.

The process is conducted by cultivating (fermenting) any of the microorganisms on an aliphatic hydrocarbon feed source, e.g., a C_1 to C_{30} n-paraffin feed in an aqueous growth medium containing available oxygen and other essential cell nutrients. If these cells are to form part of an animal feed supplement, the cells are usually rendered nonviable before use.

Usually the source of carbon will be C_6 to C_{10} light naphthas (viz., low boiling hydrocarbon oils of the C_nH_{2n+2} series and having a boiling point between 95° and 150°C.) and petroleum fractions containing them, and C_{11} to C_{30} gas oils boiling in the range of about 190° to 320°C., and petroleum fractions containing them.

The preferred n-paraffin feed for the microbes are the C_{11} to C_{30} n-paraffins. Each of the above feeds can and frequently does contain normal olefins, e.g., C_6 to C_{30} mono- and polyolefins, in varying amounts, e.g., from 0.05 to 30.0% by weight (based on total hydrocarbons in the feed).

Polycyclic aromatic compounds are usually excluded from the feed as such materials are considered to be possibly carcinogenic and could contaminate the harvested cells in feeds.

While the presence of branched aliphatic hydrocarbons (including both olefins and alkanes) in concentrations up to 30 weight percent can be tolerated in the hydrocarbon feed; concentrations in excess of 10 weight percent of non-normal aliphatic hydrocarbons are usually avoided because the microorganisms are selective preferentially to normal aliphatic hydrocarbons, especially C_{11} to C_{30} n-aliphatic hydrocarbons.

Example 1: A growth medium of the following composition was prepared.

Component	Concentration (g./l.)
n-Hexadecane*	20.0
K_2HPO_4	5.0
$(NH_4)_2HPO_4$	10.0
Na_2SO_4	0.5
$MgSO_4 \cdot 7H_2O$	0.4
$FeSO_4 \cdot H_2O$	0.02
$MnSO_4 \cdot 4H_2O$	0.02
NaCl	0.02
Water (sufficient to make a volume of 100 ml.)	

* Commercial n-hexadecane containing $<$1 weight percent C_{16} n-monoolefin.

After regulating the pH to 7.2 to 7.7, the above media was introduced into a 500 ml. Erlenmeyer flask, and the flask contents were sterilized by heating at 121°C. for 15 minutes. Then approximately 0.001 g./l. of *Brevibacterium insectiphilium* (ATCC No. 15528), previously cultured for 48 hours at 30°C. on the same medium, was inoculated into the fermentation growth medium. The fermentation media was cultured under shaking at 30°C. for 48 hours maintaining the growth pH between 6 and 7.5 throughout fermentation.

After 48 hours, the cell concentration was 6.2 g./l., and 40 weight percent of the n-hexadecane aliphatic hydrocarbon feed was utilized by the microorganism (cell yield of 77.5% based on aliphatic hydrocarbon utilized). After the completion of fermentation, the broth was centrifuged and then sterilized in a spray dryer at 180°C. for four seconds.

Example 2: Corynebacterium sp. (ATCC No. 15529) was fermented for 48 hours at 30°C. using the same growth medium and procedure set forth in Example 1 with the exception that 10 g./l. of n-hexadecane was used in this fermentation. After 48 hours, the cell growth was 9.5 g./l. and 100% of the n-hexadecane hydrocarbon feed was utilized (cell yield of 95% based on aliphatic hydrocarbon utilized).

Example 3: Corynebacterium pourometabolum (ATCC No. 15530) was grown for 48 hours at 30°C. at a pH of 6 to 7 in accordance with the procedure of Example 1 and using the growth medium thereof, but with an n-hexadecane concentration of 17 g./l. After 48 hours the cell growth was 12.8 g./l. and 100% of the n-aliphatic hydrocarbon feed was utilized (cell yield of 75% based on n-hexadecane utilized).

OTHER LIQUID FEEDSTOCKS

Hydrocarbon-Derived Products Containing Aldehydes

By the methods of *D.O. Hitzman and H.E. Alquist; U.S. Patent 3,642,578; February 15, 1972; assigned to Phillips Petroleum Co.,* proteins, amino acids, gums, and other valuable fermentation products are efficiently biosynthesized from hydrocarbon-derived feedstocks containing aldehydes which have been admixed with a nitrogen-containing compound before being passed to a fermentor.

It is known that microorganisms have the ability to manufacture edible protein by the fermentation of hydrocarbons. There is avid interest in this field for hydrocarbons represent one of the greatest sources of raw materials suitable as potential foodstuffs that can be employed to meet a continuing critical world shortage of edible protein.

Large-scale protein synthesis has not developed to the high degree of efficiency wherein high yields of protein are obtained by economical procedures.

The conversion of methane and n-paraffins to edible protein is recognized, but protein so manufactured is known to be frequently contaminated by oil and other deleterious hydrocarbons often resulting in decreased cellular yields and high production cost necessitated by extensive separating, centrifuging, and washing procedures to achieve a product free from aromatic and carcinogenic contamination and hence, suitable as foodstuffs.

Most oxidation products of methane, such as formaldehyde, are considered biocidal in nature and deleterious to the microbial fermentation process. Commercial methanol, an oxidation product of methane, which can be used as a hydrocarbon-derived feedstock often contains deleterious quantities of formaldehyde and consequently inhibits or decreases microbial productivity.

Some microorganisms are capable of incorporating and oxidizing formaldehyde but heretofore formaldehyde, even at low concentrations, was considered biocidal by those skilled in the art.

However, hydrocarbon-derived products containing substantial amounts of aldehydes, even formaldehyde, whose toxicity to microorganisms is well documented, can be successfully employed as a nutritional feedstock for microbial fermentation when they have been admixed with a nitrogen-containing compound before being passed to a fermentor.

The procedure utilizes products formed by the oxidation of hydrocarbons. High quality protein can be economically and abundantly manufactured by the oxidation and fermentation of widely available hydrocarbon sources such as natural gas, petroleum, naphtha, coal, peat, asphalt.

Hydrocarbons are oxidized and contacted with an aldehyde-reactive nitrogen-containing compound, and the water-soluble mixture formed is fed to a fermentor for microbial fermentation resulting in an uncontaminated microbial production product suitable as a protein food source.

A hydrocarbon, such as methane, can be oxidized to methanol and formaldehyde without using expensive carefully controlled oxidation reaction steps to prevent the formation of deleterious quantities of aldehydes. Time and expense are consequently jointly conserved.

Examples of those products which can be employed as microbial feedstocks include the water-soluble aliphatic alcohols, ketones, aldehydes, carboxylic acids, ethers, and polyols, preferably contain- as many as 10 carbon atoms.

Example 1: A 14 liter New Brunswick stirred fermentor suitably rigged for continuous fermentation and temperature controlled in the range of 32° to 40°C. was charged with 7 l. of suitable base medium[1] and with 500 cc of inoculum of *Pseudomonas sp. Pseudomonas methanica* NRRL 8-3449. Materials were charged to the reactor and effluent removed until the bacteria had reached an exponential rate of growth and a steady state had been reached. The following data illustrate steady-state fermentor operation (47 to 70 hours from startup).

Conditions during run	
Base medium[1]	BH6
Air input	10 l./min.
Stirrer	1,000 rpm
Growth factors	none
Feed rates	
Methanol-formaldehyde product[2]	0.1452 l./hr.
NH_4OH (22 to 26% NH_3)[3]	0.0291 l./hr.
Base medium[1]	1.200 l./hr.
Trace minerals[4]	0.016 l./hr.
Total feed rate	1.3708 l./hr.
Steady-state fermentor volume	3.5 l.
Alcohol content of effluent	1.13 %
Based on these conditions the following calculations can be made:	
Retention time in fermentor	2.55 hr.
Cell concentration (dry weight)	26.1 g./l.
Yield of dried cells/100 lbs. of methanol consumed	36.15 lbs.
Yield of dried cells/100 lbs. of methane consumed[5]	72.3 lbs.
Percent protein of cells[6]	69.4 %
Fermentor productivity[7]	10.23 g./l./hr.

[1]BH6 base medium has the following amounts of materials per liter of aqueous solution:

KH_2PO_4	2.5 g.
K_2HPO_4	2.5 g.
$(NH_4)_2SO_4$	2.0 g.
NaCl	0.1 g.
$MgSO_4 \cdot 7H_2O$	3.0 g.
$CaCl_2$	0.04 g.
Trace minerals solution[4]	10 ml.

(continued)

[2]Methanol-formaldehyde product is comprised thus:

14 parts methanol
1 part 37% aqueous HCHO

[3]NH_4OH was admixed with methanol-formaldehyde product prior to passing to the fermentor

[4]Trace mineral solution had the following amounts of the following compounds per liter of solution:

$CuSO_4{\cdot}5H_2O$	0.06 g.
KI	0.08 g.
$FeCl_3{\cdot}6H_2O$	4.80 g.
$MnSO_4{\cdot}H_2O$	0.30 g.
$Na_2MoO_4{\cdot}2H_2O$	0.20 g.
$ZnSO_4{\cdot}7H_2O$	2.00 g.
H_3BO_3	0.02 g.

[5]Assuming 100% of theoretical conversion of CH_4 to CH_3OH if methane is first oxidized to methanol.

[6]Percent protein equals percent N x 6.25.

[7]Fermentor productivity is in grams of dried cells per liter of ferment per hour retention time in fermentor.

Example 1 clearly exemplifies the efficient productivity and high-cellular protein content achieved by the process.

Example 2: A run was effected as in Example 1 except that ammonium hydroxide was charged to the fermentor separately without prior admixing with the methanol-formaldehyde feedstock. The culture was eradicated and fermentation ceased.

Example 2 demonstrates the criticality of admixing the nitrogen-containing compound to the oxidized hydrocarbon-containing feedstock prior to passing the feedstock to the fermentor.

Oxygenated Hydrocarbons

The process of *J.D. Douros, Jr., L.A. Naslund and W.J. Lahl; U.S. Patent 3,546,071; December 8, 1970; assigned to Esso Research and Engineering Company* requires the use of oxygenated hydrocarbons particularly water-soluble, oxygenated hydrocarbons as a primary source of carbon.

Hydrocarbon fractions, such as petroleum liquids and gases, are oxidized, or partially oxidized, and then contacted with a microorganism under fermentation conditions in the presence of cellulose to produce high yields of protein. The process may be carried out continuously or batchwise.

Composition for the fermentation media particularly for bacteria at the outset of fermentation is as follows.

Component	Concentration (grams per liter) Can Use	Usually Use	Preferably Use
Oxygenated hydrocarbons	4 - 120	5 - 80	10 - 50
Cellulose*	2 - 60	2.5 - 40	5 - 25
K_2HPO_4	0.5 - 15	1 - 10	2 - 8
$(NH_4)_2HPO_4$	5 - 15	7 - 13	8 - 12
Na_2SO_4	0.1 - 1.0	0.2 - 0.9	0.3 - 0.8
$FeSO_4 \cdot 7H_2O$	0.002 - 0.5	0.005 - 0.04	0.01 - 0.03
$MgSO_4 \cdot 7H_2O$	0.1 - 0.7	0.2 - 0.6	0.3 - 0.5
$MnSO_4 \cdot 7H_2O$	0.002 - 0.05	0.005 - 0.04	0.01 - 0.03
NaCl	0.002 - 0.05	0.005 - 0.04	0.01 - 0.03
Water	Remainder to equal 100 weight percent		

*Cellulose is preferably in strips such as about 1/8 to 2 in. wide and ½ to 6 in. long. Desirable strips are ¼ in. wide and 1 in. long.

The process employs a crude oxygenated fraction such as a crude fraction having the following composition.

	Percent
Methyl alcohol	56.6
Acetone	29.0
Ethyl alcohol	8.1
Ethyl acetate	2.7
Methyl acetate	2.0
Dimethyl propane	0.3
t-Butyl alcohol	0.3
Methylethyl ketone	0.3
Methylisopropenyl ketone	0.3
H_2O	0.3

The cellulose is preferably cellulose acetate strips which are prepared by treating wood pulp with acetic acid, acetic anhydride and sulfuric acid as a catalyst. The cellulose is fully acetylated (three acetate groups per glucose unit) and at the same time the sulfuric acid causes degradation of the cellulose polymer so that the product contains only about 200 to 300 glucose units per polymer chain. At this point in the process the cellulose acetate is partially hydrolyzed by the addition of water until an average of 2 to 2.5 acetate groups per glucose unit remain. This product is a thermoplastic.

Unexpected, desirable results are secured by the addition of the cellulose strips to the fermentation process in that the growth rate and the conversion rate of the oxygenated hydrocarbon is markedly increased.

Example: Micrococcus cerificans (14987) was grown and cellulose strips were added in addition to the various oxygenated hydrocarbons. The results secured are listed in the following table.

Micrococcus cerificans **Grown on Various Oxygenated Hydrocarbons**

Compound*	g./l. added	Cell wt. g./l. without cellulose strips	g./l. cellulose added	Cell wt. g./l. with cellulose strips	Percent increase with cellulose strips
Cetylalcohol	7.3	5.58	5	5.75	3
Ethyl alcohol	11.0	3.69	5	4.40	19
Sec-butyl alcohol	8.9	1.11	5	1.41	27
Oleic acid	7.5	5.76	5	7.91	37
Methyl alcohol	15.4	4.62	5	7.63	60
Isopropyl alcohol	9.6	1.04	5	4.62	444

*All compounds added on an equal carbon basis (480 mmoles).

From the foregoing data it is evident that cell growth rates and oxygenated hydrocarbon conversion rates can be greatly increased by the use of cellulose strips. Cell weights are increased from 3 to 444% depending on the oxygenated hydrocarbon used. No growth is observed with cellulose and *Micrococcus cerificans* without an oxygenated hydrocarbon carbon source.

Growth on an Extract of Coal

Coal is extracted with water, and the extract is employed as nutrient for microorganisms in the method of *M.J. Rose, J.M. Carosella, J.D. Corrick and J.A. Sutton; U.S. Patent 3,540,983; November 17, 1970; assigned to the U.S. Secretary of the Interior.*

Only very minor amounts of the coal constituents need be dissolved in the water during extraction. Extracts containing less than 0.25 weight percent of the coal will cultivate the growth of any microorganisms which can employ coal as nutrient.

Extraction is accomplished by contacting the coal with fresh water (e.g., distilled or tap water) in the usual extraction manner. A single contacting step or countercurrent contact is suitable. The same portion of water can be passed in contact with the coal a plurality of times, each contacting step being carried out at the same or a different temperature.

Alternatively, the coal can be contacted with a plurality of separate water portions, each resultant extract serving as nutrient medium.

Contacting time depends upon such factors as the solids:liquid ratio, particle size of the coal, and the extractant water temperature. Such temperatures can vary from about 25° to 125°C. Coal particle size should not be so small as to render solids-liquid separation too difficult and should not be so large as to unduly increase the contact time required. A coal particle size ranging from about 8 mesh to 200 mesh, and a coal:water weight ratio of about 1:1 to 1:100 are suitable for the purpose of this process.

After separating the water extract from residual solids by, for example, filtration or centrifugal force, the extract is sterilized by heating, under pressure, to a temperature of about 121° to 125°C.

Upon inoculation of the sterilized extract with appropriate microorganisms, environmental conditions are adjusted with regard to the organism's optimum growth temperature range, and with regard to whether the organism is aerobic, anaerobic or facultative.

As an example, the yeast *Candida tropicalis* grows rapidly in the presence of a water extract of lignite, leonardite or sub-bituminous coal when the following environmental growth conditions are employed: (1) a temperature of about 20° to 35°C., preferably about 30°C.; and (2) a pH of 4 to 8 in the presence or absence of air, preferably in the presence of air at a pH of 4 to 5. Aerobic conditions are established by, for example, bubbling air through the nutrient medium, or by stirring.

Although the water extract of coal is capable of supporting growth of microorganisms without the addition of supplementary growth materials, additional nutrients can be added if desired.

Example 1: Lignite was ground to -200 mesh in a ball mill. 100 grams of the comminuted lignite was then mixed with 1 liter of cold tap water and stirred at 100 to 300 rpm for four hours. Thereafter the mixture was filtered through Whatman #5 filter paper and once through a Seitz filter. Resultant filtrate was sterilized at 121°C. and 15 psig for 15 minutes.

A *Candida tropicalis* culture was injected into the sterilized water extract and into a sterilized tap water control sample (both at 30°C.) to establish an initial viable cell count of 100/ml. At the end of five days the control sample had a reduced viable cell count of 70/ml. while the water extract had an increased viable cell count of 760,000/ml.

Example 2: This example illustrates the effects of serially extracting a body of coal with separate water samples.

A 100 gram portion of -200 mesh leonardite was contacted for two hours with 1 liter of distilled water at 100°C. Leonardite residue from this first extraction was then contacted with another liter of distilled water for two hours to obtain a second water extract. In this manner, six fractions of water extract were prepared from the same portion of leonardite.

While the first extract contained approximately 3% (by weight) of the leonardite dissolved therein, and the sixth extract contained a little more than 0.5% of the original leonardite dissolved therein, infrared spectrometry showed only minor differences in the spectra of untreated leonardite and the several extracts.

Each sterilized extract (along with a control sample of distilled water) was then inoculated with *C. tropicalis* to establish a viable cell count of 500/ml. No supplemental nutrients were employed. The cell count in each sample (maintained at 30°C.) was taken once a day for several days, and the following results were obtained.

	Approximate Viable Cell Count per Milliliter				
Time Lapse in Days	Control	First Extract	Second Extract	Fourth Extract	Sixth Extract
1	600	500	500	525	500
2	650	3,500	3,000	3,500	1,500
3	600	3,000	2,500	5,000	3,000
4	300	70,000	55,000	65,000	45,000
5	150	90,000	80,000	120,000	100,000
6	35	95,000	130,000	80,000	50,000

This table shows that the sixth extract, despite the fact that it had considerably less leonardite material dissolved therein in comparison to the first extract, supported growth about as well as the first extract.

Additional extracts made from the same leonardite sample as used in this example were capable of supporting growth to the same degree as the first six extracts, although some of the additional extracts contained less than 0.25 weight percent of the original leonardite dissolved therein. However, the growth rate was slower.

Other tests with leonardite water extracts have shown that concentrating the extract (prior to inoculation) by partial evaporation of the water therein does not increase microorganism growth. On the contrary, although the final yield equals that obtained with unconcentrated extract, it takes several days longer to reach optimum growth.

Protein from Cyclohexane Oxidation Products

W. Griehl and C. Suter-Homuth; U.S. Patent 3,523,064; August 4, 1970; assigned to Inventa AG für Forschung und Patentverwertung, Switzerland describe a process for the production of cellular protein from the waste liquors obtained in the cyclohexane oxidation. Microorganisms of the family of Pseudomonadaceae convert these waste products to proteins in alkaline medium at ambient temperatures at 24° to 30°C. and pH levels of substantially 6.7 to 9.2 within 50 to 90 hours. The process has the twofold advantage of manufacturing a useful product from wastes and to prevent water pollution.

In order to convert the organic waste products into protein, they are subjected to fermentation, either in the form as originally obtained or after acidulation, e.g., with sulfuric acid, and separation therefrom as upper and lower layers, respectively, or else after extraction with cyclohexane and ensuing distillation.

It is also feasible to totally extract, e.g., with methylethyl ketone, from the lower layer after acidulation, all organic acids present. The wastes usually are employed in concentrations ranging from 0.2 to 50 weight percent, calculated on the nutrient medium.

Example: Pseudomonas fluorescens was shaken at 25°C. for 48 hours in a wide neck Erlenmeyer flask with a culture medium as described below. The medium previously had been sterilized for 30 minutes at a pressure of 0.7 kg./cm.^{2} at 110°C, and consisted of an aqueous solution containing:

$FeSO_4 \cdot 7H_2O$	0.01%
$MgCl_2 \cdot 6H_2O$	0.04%
$MgSO_4 \cdot 7H_2O$	0.05%
$NaNH_4HPO_4 \cdot 4H_2O$	0.35%

(continued)

K_2HPO_4 anh.	1.00%
Mixed monocarboxylic acids (C_4 to C_6 in proportions of 1:8:3)	0.20%

The pH of the nutrient solution was 7.0.

A nutrient medium, sterilized as described above, was seeded with this culture medium in the proportion of one part of seed culture to each 10 parts of nutrient medium. The nutrient medium had the following composition:

$MgSO_4 \cdot 7H_2O$	0.05%
K_2HPO_4 anh.	1.00%
$NaNH_4HPO_4 \cdot 4H_2O$	2.00%
NaCl	0.05%
Water	Balance

The wastes of the cyclohexane oxidation, in the form described above as the lower layer after acidulation, i.e., an aqueous solution of the sodium salts of dicarboxylic acids, succinic and glutaric acid, monocarboxylic acids, formic, butyric, valeric and caproic acid, and free NaOH, were adjusted to a pH of 7 with 10% H_2SO_4 in proportions of 1:2, calculated on the fermentation solution, and were added to the combined media.

Conversion into bacterial protein was carried out at 25°C. with agitation of 450 rpm and aeration with ½ volume O_2 per volume reaction mixture per minute. The pH value, tending to rise, was held at 7.5 by addition of 10% H_2SO_4. Silicone oil was added as required.

The "lower layer" solution was dropped in continuously, and the reaction mixture, after 70 hours, had a microorganism propagation factor of 100 to 120.

Upon conversion of the waste into bacterial dry substance, 100 parts carbon compounds yielded 50 parts dry bacterial substance having a protein proportion of 50%.

The fermentation solution, after 70 hours, was centrifuged at 4,500 rpm. After air drying at room temperature, the cellular sediment then was post-dried to constant weight at 60° to 65°C. When the intracellular amino acids were to be isolated and identified, they were acid digested and adsorbed on ion exchangers in the conventional manner and identified chromatographically.

Mixture of Aliphatic Mono- and Dicarboxylic Acids

J.W. Woldendorp; U.S. Patent 3,619,368; November 9, 1971; assigned to Stamicarbon NV, Netherlands has also described fermentations in which a technically attractive production of protein-rich cellular material in high yields is possible by the use of a culture medium which contains a carbon source consisting of a mixture of aliphatic mono- and dicarboxylic acids with fewer than eight carbon atoms.

In practice, a carbon source like this becomes available in fairly large quantities and very cheaply in the wash water obtained when, in the preparation of cyclohexanone from cyclohexane or cyclohexanol, the crude final product is washed with water and/or a dilute sodium hydroxide solution.

With the use of a culture medium containing a carbon source in the form of aliphatic mono- and dicarboxylic acids excellent results are obtainable if during the cultivation of protein-rich *Eumycetes* the pH of the liquid culture medium is maintained at a value lying between 5.5 and 8. Since the acid concentration in the culture medium decreases during the growth of the microorganism, it is necessary, in order to maintain the pH at the desired value, to add a pH reducer during this growth, e.g., sulfuric acid or a similar acid, or to use a buffered culture medium.

In connection with the ever-growing production of cyclohexane, the process is of great practical importance for the processing of the wash water obtained in this production. In practice, this wash water may in most cases be drained only after a costly purification, e.g., by biological breakdown of the organic acids contained in it. This process not only purifies the wash water, but also yields a valuable product without any previous purification being required.

When the process was carried out as a continuous process, it was furthermore found that with addition to the culture medium of a small amount of carbon source easily attacked by the microorganism concerned, for instance glucose, if the microorganism is a *Candida* strain or acetic acid in the case of a *Trichosporon* strain, a much higher concentration of mono- and dicarboxylic acids can be used than in the case of a discontinuous realization.

In a continuous process, excellent results can be obtained if per gram of carbon contained in the mono- and dicarboxylic carbon source an amount of 1 to 10 milligrams of carbon in the form of an easily accessible carbon source is added to the culture medium. The mono- and dicarboxylic carbon source can then be continuously supplied to the culture medium in a concentration of 25 to 30 grams per liter calculated as carbon.

Example 1: A strain of *Candida brumptii* was seeded in malt extract and then incubated for 18 hours at 30°C. in a shaking incubator.

Of the active culture obtained in this way, 0.2 ml. was seeded in a 300 ml. flask into which 50 ml. of a liquid culture medium had previously been introduced, which medium had a pH of 6.5 and possessed the following composition (in grams per liter):

Adipic acid	1.00
Butyric acid	0.50
Valeric acid	0.74
Hydroxycaproic acid	0.76
Acetic acid	0.10
Ammonium sulfate	3.00
Primary potassium sulfate	7.00
Magnesium sulfate·$7H_2O$	0.40
Yeast extract	0.10

Thereafter, incubation at 30°C. was applied for 24 hours, and next the resulting cellular material was isolated by centrifugation washed and dried.

The yield was 1.5 grams of yeast cells per liter of culture medium. The crude protein content of the yeast was 54% by weight.

Example 2: A strain of *Candida lipolytica* was precultivated in malt extract in the same way as described in Example 1.

10 ml. of the resulting active culture was seeded in a 1.5 liter fermentation vessel containing 1 liter of a liquid culture medium obtained by adding 3 g. of ammonium sulfate, 1 g. of monopotassium phosphate, 0.4 g. of magnesium sulfate ·$7H_2O$ and 1 ml. of cornsteep liquor to 50 ml. of washing water obtained in the preparation of cyclohexanone, and making up with water to 1 liter.

The washing water contained 1.44 g. eq. of acid per liter, composed as follows (in gram equivalents):

0.114 of formic acid
0.054 of acetic acid
0.028 of propionic acid

(continued)

0.088 of butyric acid
0.308 of valeric acid
0.068 of γ-valerolactone
0.076 of caproic acid
0.158 of ϵ-hydroxycaproic acid
0.028 of succinic acid
0.010 of oxalic acid
0.080 of glutaric acid
0.262 of adipic acid

After seeding, the liquid was stirred in the fermentation vessel for 24 hours at 28°C. by means of a turbine stirrer (1,200 rpm) and aerated with 60 liters of air per hour, the pH being meanwhile kept at 6.5 by continuous addition of sulfuric acid. The resulting cellular material was then isolated by centrifuging, washed and dried.

The yield was 2.4 g. of dried cellular material having a crude protein content of 54% by weight. Of the original amount of carbon present in the organic acids, 84% proved to have been converted.

Example 3: A strain of *Candida tropicalis* was precultivated in malt extract. 0.2 ml. of the resulting active culture was seeded in a 300-ml. flask containing 50 ml. of a liquid culture medium of pH 5.5, having the following composition (in grams per liter):

Acetic acid	0.27
Propionic acid	0.11
Butyric acid	0.18
Valeric acid	0.05
Oxalic acid	0.74
Succinic acid	0.15
Glutaric acid	0.40
Adipic acid	1.42
ϵ-hydroxycaproic acid	0.56
Ammonium sulfate	3.00
Monopotassium sulfate	1.00
Magnesium sulfate	0.20
Yeast extract	0.50

After 48 hours' cultivation in a shaking incubator at 30°C., an amount of 5 ml. of the resulting culture medium was seeded in 50 ml. of culture medium of the same original composition as given above. After 36 hours' cultivation at 30°C. this gave 1.2 g. of dry yeast per liter of culture medium.

Gaseous and Solid Hydrocarbon Feedstocks

GASEOUS HYDROCARBONS

Butane

K. Tanaka, K. Ohshima, K. Kimura and M. Yamamoto; U.S. Patent 3,639,210; February 1, 1972; assigned to Kyowa Hakko Kogyo Co., Ltd., Japan have directed their attention to the utilization of gaseous hydrocarbons, particularly butane, which are the most inexpensive of the hydrocarbons and which can be supplied in abundance.

It has been found that hydrocarbon assimilating strains of microorganisms, especially those microorganisms belonging to the genus Brevibacterium are excellent in the utilizability of gaseous hydrocarbons such as butane as the main carbon source in the fermentation medium.

It has been found that when strains such as Brevibacterium are used, various useful fermentation products, such as for example, amino acids, organic acids, saccharides, nucleic acids, vitamins and cells of microorganisms, are produced when using gaseous hydrocarbons as the main carbon source.

The advantages of this fermentation as compared with conventional fermentation processes using saccharides as starting materials and those recently developed fermentation processes using liquid hydrocarbons as the carbon source in the fermentation medium are listed on the following page.

(1) The gaseous hydrocarbons are abundant and inexpensive.

(2) The storage, transportation and feeding of the gaseous hydrocarbons to the culture medium are relatively simple.

(3) The recovery of the residual gaseous hydrocarbons from the fermentation medium can be readily effected.

(4) The control and management of the concentration of the gaseous hydrocarbon source during fermentation is readily accomplished.

(5) The fermentation liquor after completion of the fermentation is clear, and thus separation thereof from residual starting materials and purification thereof are readily effected.

In fermentations using gaseous hydrocarbons as the main carbon source, it is also possible to directly apply various fermentation procedures normally employed in conventional fermentations using saccharides or liquid hydrocarbons.

However, in order to effectively increase the proportions of gaseous hydrocarbons dissolved in the fermentation medium, small amounts of nonutilizable liquid hydrocarbons or a substance similar in effect thereto and a surface active agent are added to the fermentation medium, whereby the fermentation rate can be substantially enhanced.

Microorganisms which can be used in this process are the hydrocarbon-assimilable strains, particularly those belonging to the genus Brevibacterium. The method of this process is particularly applicable to the microorganism *Brevibacterium ketoglutamicum,* ATCC No. 15,587.

As for the fermentation per se, either a synthetic culture medium or a natural nutrient medium is suitable as long as it contains the essential nutrients for the growth of the microorganism strain employed. Such nutrients are well known in the art and include substances such as a carbon source, a nitrogen source, inorganic compounds and the like which are utilized by the microorganism employed in appropriate amounts.

A hydrocarbon gas such as methane, ethane, propane, n-butane, isobutane, propylene, butylene, and the like, as well as mixtures of these materials, is used in the fermentation process as the main carbon source. Small amounts of other carbon sources may also be present in the fermentation medium, such as for example, liquid hydrocarbons including straight and branched-chain paraffins (alkanes) having from 5 to 24 carbon atoms, such as n-pentane, n-octane, n-decane, etc., carbohydrates such as glucose, fructose, maltose, sucrose, starch, starch hydrolyzates, molasses, etc., or any other suitable carbon source such as glycerol, mannitol, sorbitol, organic acids, glutamic acid, etc. These substances may be used either alone or in mixtures of two or more but in small amounts when compared to the primary source of carbon, that is the hydrocarbon gases mentioned above.

As a nitrogen source, various kinds of inorganic or organic salts or compounds, such as urea, ammonia or ammonium salts such as ammonium chloride, ammonium sulfate, ammonium nitrate, ammonium phosphate, etc., or one or more than one amino acid mixed in combination, or natural substances containing nitrogen, such as corn steep liquor, yeast extract, meat extract, peptone, fish meal, bouillon, N-Z Amine (a series of casein hydrolyzates), casamino acid, fish solubles, rice bran extract, ribonucleic acid, etc., may be employed.

Again these substances may be used either singly or in combinations of two or more. Inorganic compounds which may be added to the culture medium include magnesium sulfate, sodium phosphate, potassium dihydrogen phosphate, potassium monohydrogen phosphate, iron sulfate or other iron salts, manganese chloride, calcium chloride, etc.

Example: 20 ml. of the following culture medium is prepared in a 250 ml. triangular flask:

2% NH_4NO_3
0.05% $Na_2HPO_4 \cdot 12H_2O$
0.05% KH_2PO_4
0.01% $MgSO_4 \cdot 7H_2O$
0.001% $MnSO_4 \cdot 4H_2O$
0.001% $FeSO_4 \cdot 7H_2O$
0.001% $ZnSO_4 \cdot 7H_2O$
0.001% $CaCl_2 \cdot 2H_2O$
50γ/l. $CuSO_4 \cdot 5H_2O$
10γ/l. H_3BO_3
10γ/l. $Na_2MoO_4 \cdot 2H_2O$
0.1% corn steep liquor

The pH of this medium is 7.0.

Brevibacterium ketoglutamicum 15,587, previously cultured with aerobic shaking in a bouillon medium, is inoculated into the culture medium and the medium is subjected to shaking at a temperature of 30°C. by means of a thermostatic stirrer. Culturing is conducted by supplying to the culture medium a mixed gas of n-butane and air having a butane gas concentration in the gas mixture of 50% by volume. Unused butane gas can be recycled and fresh butane gas can be added to the culture medium such that the butane concentration in the culture medium is always about 50% by volume.

On the 3rd day of the cultivation, 0.1% of cetyltrimethylammonium bromide is added and the cultivation is continued for an additional 4 days. After this period, 0.30 g./dl. of L-glutamic acid, 0.12 g./dl. of alanine, 0.05 g./dl. of valine, and 0.001 g./dl. of lysine are produced in the culture medium.

The same cultivation is effected using 50 triangular flasks of 250 ml. in volume, and the liquors after cultivation are united and separated from the cells of the microorganisms by centrifugal separation to form a cultivation completed liquor.

One liter of this liquor is adjusted to a pH of 2.0 by the addition of hydrochloric acid, and is adsorbed on the cation exchange resin. Amberlite IR 120 (H-type), and the individual amino acids are separated and eluted by successively elevating the pH. After concentration, 2.2 g. of L-glutamic acid, 0.8 g. of alanine, 0.4 g. of valine and 5 mg. of lysine are obtained. In addition, 4 g. of dry cells of the microorganism are recovered.

Propane and Butane in Natural Gas Fractions

G. Orgel, E.W. Pietrusza and G.G. Joris; U.S. Patent 3,622;465; November 23, 1971; assigned to Allied Chemical Corporation have described an efficient biosynthetic process for converting straight chain hydrocarbons into edible protein.

In the process an inoculum of a strain of the microorganism *Arthrobacter simplex* is introduced into an aqueous culture medium containing a C_3 to C_{18} straight chain hydrocarbon as the principal source of assimilable carbon and energy, an assimilable source of nitrogen, and essential inorganic salts.

A hydrocarbon feedstock suitable for use in the process is one which is capable of releasing carbon in a form suitable for metabolic utilization by the microorganism. C_3 to C_{18} straight chain alkanes and alkenes have been found to be suitable for Strain B-129 and C_6 to C_{18} straight chain alkanes and alkenes for *A. simplex* ATCC No. 6946. Although the presence of branched aliphatic hydrocarbons (including both olefins and alkanes) in concentrations up to 30 weight percent can be tolerated in the hydrocarbon feed without poisoning the microorganism, concentrations in excess of 10 weight percent of branched chain aliphatic hydrocarbons in conjunction with the straight chain hydrocarbon feedstock are usually avoided. The microorganisms will preferentially assimilate normal aliphatic hydrocarbons. The preferred procedure uses a natural gas fraction comprising in substantial proportions, propane or butane or mixtures thereof as the principal source of assimilable carbon and energy, and the B-129 strain of *A. simplex* as the microorganism grown on this feedstock.

The culturing of the microorganisms in accordance with the process is carried out under aerobic conditions. While sufficient aeration is generally obtainable in small flasks by mechanical agitation during the fermentation period, it is important in larger scale operations that the culture broth be stirred mechanically and that sterile air be introduced into the vessels by known methods.

It has been generally found desirable to introduce from about one-half to one volume of air per minute for each volume of fermentation broth. If a gaseous hydrocarbon feed is used, a ratio of air/hydrocarbon by volume of from 1/1 to about 4/1 is suitable. Alternatively, pure oxygen, oxygen enriched air, or a mixture of oxygen, nitrogen and carbon dioxide may be used to maintain aerobic conditions.

Example 1: A culture of Strain B-129 of *Arthrobacter simplex* developed by the enrichment-of-culture technique was maintained in an aqueous growth medium under a 50/50 v./v. percent propane/air atmosphere. The liquid medium was of the following composition:

NH_4NO_3	1.0 g.
$MgSO_4 \cdot H_2O$	0.20 g.
$CaCl_2$	0.20 g.
KH_2PO_4	1.00 g.
K_2HPO_4	1.00 g.
Deionized water	1,000 ml.
*Trace element solution	1.0 ml.

*Trace element solution contains:

H_3BO_3	0.0025 g./ml.
$MnCl_2 \cdot 4H_2O$	0.0015 g./ml.
$ZnCl_2$	0.0001 g./ml.
$CuCl_2 \cdot H_2O$	0.0005 g./ml.
MoO_3	0.0005 g./ml.
Sequestrine 330 Iron Chelate	0.001 g./ml.

The medium was dispersed into side arm Erlenmeyer flasks, fitted with a gassing stopper. Medium volumes were as follows for different flask sizes: 50-ml. medium/250-ml. flask and 100-ml. medium/500-milliliter flask. The liquid:gas ratio was 1:5 in each case.

This liquid growth medium was inoculated with a 1 volume percent inoculum. The pH of the medium at inoculation ranged from 6.7 to 6.9. After inoculation the flasks were evacuated and refilled with the propane/air mixture to atmospheric pressure.

The flasks were then incubated at room temperature (24° to 26°C.) on a reciprocative shaker at 115, 2 1/2-inch strokes/minute or on a thermostated gyrotory shaker (30°C./250 rpm). Propane and oxygen were consumed, carbon dioxide was produced and the turbidity of the growth medium increased steadily. All of which indicated that the organism was proliferating.

Example 2: A cultivation of Strain B-129 was carried out in a large fermentor using the growth medium of Example 1. Experimental conditions were as follows: liquid medium volume, 10 liters; inoculum concentration, 5.0%; agitation, 400 rpm; gas flow, 5 liters per hour (50/50 air/propane, v./v.); initial pH, 6.5, maintained at 6.5 by periodic addition of KOH; temperature, 29°C.; time, 66 hours; cell harvest, continuous centrifugation in Sorval Hi-Speed Centrifuge, 4.5 l./hr. flow rate, 17,000 rpm at 2°C., yield, 6.03 g.

The cells were washed 3 times with deionized cold H_2O and lyophilized. The lyophilized cells were a white to light buff color and had a fluffy texture. Typical analysis of lyophilized *Arthrobacter simplex* B-129:

%N = 11.94-13.89 (3 preps.)
%C = 46.33-46.0
%H = 6.77-7.2
%S = 1.5
% Protein Content* = 74.63-86.81

*Assuming all nitrogen is proteinaceous.

Methane and Natural Gas

B. Wolnak, B. Andreen and C.-Y. Wen; U.S. Patent 3,649,459; March 14, 1972; assigned to Institute of Gas Technology had isolated and characterized a mixed culture of microorganisms which are capable of oxidizing methane and natural gas in a nutrient medium containing only inorganic salts to produce cellular material high in protein and vitamin content. The mixed culture contains a gram-negative rod about 0.5 to 3.0μ wide and 2 to 10μ long as the dominant organism, several smaller bacteria and an occasional yeast cell. The mixed culture can be isolated from the plant Elodea or a soil sample.

The culture, when fed methane or natural gas, produces unexpectedly high yields of valuable proteins and vitamins. The cellular material produced can be used as a food and feed supplement and potentially could serve as food material for large percentages of the world's population.

In addition, the cellular material is unexpectedly rich in valuable vitamins which can be isolated and sold in pure form or can be used as part of the cellular food supplement material.

Example: Two runs, each of 50 ml. of the growing culture was added to 450 ml. of the medium shown in Table 1 and contacted with gas containing 40% by volume natural gas, 40% oxygen, 5% carbon dioxide and 15% nitrogen. Gas flow rate was maintained at 50 ml./minute for about 120 hours.

The temperature was maintained at 23° to 25°C., the pH 6.8 to 7.2. The cells obtained from the runs were isolated by centrifugation at 10,000 rpm for 20 to 30 minutes. The cells from the two runs were combined to provide a single sample. The average of the gross analysis of the cell composition of this sample is shown in Table 2.

The protein content is somewhat lower than expected in a cell of this type, but this may be due to the presence of nitrate in the moisture retained on the surface of the cell or to the fact that the 6.25 factor may be too low.

The fat content is also low but not inordinately so. Samples of the cells were stained to determine some of the characteristics of the fat in them. On a qualitative basis, the cells have a low fat content. The ash content was surprisingly low considering the fact that the organism was grown in a high ash content medium.

Analysis for six amino acids showed the presence of 1.08% tryptophane, 3.30% lysine, 1.42% histidine, 0.91% methionine, 9.12% leucine and 5.15% glutamic acid.

TABLE 1: SALTS IN MEDIA (g./500 ml.)

Salt	Grams	Micrograms
$NaNO_3$	1.0	1,000,000
$MgSO_4 \cdot 7H_2O$	0.1	100,000
K_2HPO_4	0.045	45,000
$CaSO_4$	0.01	10,000
$FeSO_4 \cdot 7H_2O$	0.001	1,000
Na_2HPO_4	0.0002	200
$Co(NO_3)_2 \cdot 6H_2O$	0.0002	200
H_3BO_3	0.0002	200
$MnSO_4$	0.00004	40

TABLE 2

Composition	Percent by Weight (dry basis)
Total N	8.24
Nonprotein N	2.8
Protein N	5.44
Fat	4.0
Ash	2.6
Protein (N x 6.25%)	34.0
Carbohydrate (calculated)	59.4

Solids content of cells as isolated equals 5.2%.

Liquefied Petroleum Gas

The method of *O. Imada, K. Hoshiai and M. Tanaka; U.S. Patent 3,635,796; January 18, 1972; assigned to Kyowa Hakko Co., Ltd., Japan* involves a fermentation process for culturing microorganisms using a liquefied hydrocarbon gas which is hereinafter referred to as "LPG" (liquefied petroleum gas).

The process is concerned with culturing microorganisms where the supply and dissolution of LPG in the culturing process is based on the rate of LPG consumption by the microorganisms as well as the maintenance of a minimum amount of supplied and dissolved LPG which will not lower the rate of propagation of the microorganisms or formation of the product.

Liquefied hydrocarbon generally has a very low solubility in water and the volatility thereof under ordinary fermentation conditions such as for example, temperature of 20° to 40°C. and atmospheric pressure is also low. Thus, there is no dissipation of the liquefied hydrocarbon toward the outside of the system as a result of the aeration necessary for aerobic culturing. Even if the liquefied hydrocarbon is originally in a state where oil and water are in two separate phases, the transfer and dissolution of the liquefied hydrocarbon from the oil phase into the water phase takes place as the oily substance which is dissolved in the water phase is consumed.

Thus, it is not difficult to supply the liquefied hydrocarbon (oily substance) in a state utilizable for the microorganisms and there is not any peculiarity distinguishable from the conventional and general fermentation method.

However, the situation is different with LPG. The expression "LPG" refers to a gaseous hydrocarbon which is not liquefied at ordinary temperatures and atmospheric pressure. LPG includes liquefied hydrocarbon gases such as methane, ethane, propane, n-butane, isobutane, propylene, butylene, and the like.

LPG has a very low solubility in water and is difficult to dissolve into the culturing liquor prior to culturing in an amount sufficient for utilization by the microorganisms as a carbon source. Even if the solubility of the LPG in the culturing liquor is increased by compression or cooling, the dissolved LPG would be rapidly vaporized when the aeration is conducted under the ordinary culturing conditions such as ordinary culturing temperatures and atmospheric pressure.

Also, the solubility becomes reduced to the concentration in equilibrium with the partial pressure of LPG in the gaseous phase. Almost all of the LPG is dissipated toward the outside of the culturing system and thus is not utilizable for the microorganisms. Consequently, aerobic fermentation using LPG as a raw material has been practically impossible in conventional culturing systems.

It is possible to provide for the consumption of LPG by the microorganisms by blowing LPG into the culturing liquor and adjusting the LPG concentration of the culturing liquor. Thus, the process for culturing a microorganism in an aqueous nutrient medium under aerobic conditions in the presence of a liquefied hydrocarbon gas as the main carbon source comprises adding fresh liquefied hydrocarbon gas to the culture medium, determining the concentration of dissolved liquefied hydrocarbon gas in the culture medium and controlling the addition of fresh liquefied hydrocarbon gas depending upon the detected concentration of the liquefied hydrocarbon gas in the culture medium, the supply and dissolution of liquefied hydrocarbon gas being based on the rate of liquefied hydrocarbon gas consumed by the microorganisms.

Alternatively, the partial pressure of the liquefied hydrocarbon gas in the vent gas can be measured and used to control the rate of liquefied hydrocarbon gas supplied to the culture medium. The principle of the process is based on the supply and dissolution of LPG according to the rate of LPG consumption by the microorganisms, which changes as the culturing process proceeds, as well as the maintenance of the concentration of supplied and dissolved LPG to a minimum level which will not lower the rate of propagation of the microorganisms or the rate of formation of the product.

By controlling the rate of LPG supply according to the rate of LPG consumption by the microorganisms and by adjusting the concentration of dissolved LPG, the necessary LPG for culturing can be sufficiently and economically supplied to the fermentation medium and the fermentation can be smoothly accomplished.

Example: To a fermentation tank which is 6 m. high and 2 m. in diameter is supplied 10 m.^{3} of an aqueous solution containing 0.05% of KH_2PO_4, 0.05% of K_2HPO_4, 0.05% of $MgSO_4 \cdot 7H_2O$, 0.01% of $MnSO_4 \cdot 4H_2O$, 0.01% of $FeSO_4 \cdot 7H_2O$, 0.30% of $(NH_4)_2SO_4$, 0.5% of yeast extract, and 10 mg./l. of thiamine.

The solution is heated to 110°C. for 5 minutes for sterilization and aerated with 1 to 3 Nm.^{3}/min. of sterile air. The temperature is maintained at 28°C. by passing cooling water through a cooling jacket. A device for supplying NH_3 gas by means of a glass electrode pH controller is provided to keep the pH at about 5.8 to 6.0 during culturing. Into the solution is inoculated 300 l. of a prepropagated solution of *Candida rigida* No. 113 (butane-assimilable microorganism).

The operation is started with a stirring revolution rate of 250/min. The LPG flow rate control valve is adjusted so that the partial pressure of butane gas may be kept to 0.05% or less in the vent gas. After about 45 hours of culturing, the propagation of cells of the microorganisms reaches a maximum, and culturing solution containing 25 g./l. of the cells of microorganism is obtained.

The total amount of butane supplied is 1,020 kg. and butane loss to the outside of the system is about 60 kg. Thus the utilization efficiency is 94.1%. The efficiency of assimilation into the cells of the microorganism is 26.8% and the total yield is 25.3%.

In similar culturing, where the culturing is conducted by adjusting the partial pressure of butane gas in the vent gas to 0.1%, the amount of propagated microorganism cells reaches 29 g./l. after about 40 hours of culturing, and the total amount of the supplied butane is 1,105 kg. and the butane loss to the outside of the system is 90 kg.

Thus, the utilization efficiency is 91.9%. The efficiency of assimilation into the cells of the microorganisms is 29.9%, and the total yield is 27.0%.

When the culturing is conducted under the culturing conditions without adjusting the partial pressure of butane gas in the vent gas but controlling the flow rate of supplied gas to about 22 to 25 kg./hr., the amount of the propagated cells of microorganism reaches 19.5 grams per liter after 45 hours of culturing.

The total amount of the supplied butane is 1,150 kg. and the butane loss to the outside of the system is 270 kg. The utilization efficiency of the butane is 76.5%, and the efficiency of the assimilation into the cells of the microorganisms is 22.8%. Yield: 17.5%.

In a continuation of this work *O. Imada, K. Hoshiai and M. Tanaka; U.S. Patent 3,645,846; February 29, 1972; assigned to Kyowa Hakko Kogyo Co., Ltd., Japan* describe a process in which LPG can be supplied to the culture medium without excess or shortage and the loss of LPG can be substantially prevented.

LPG has a very low solubility in water, its solubility being almost proportional to the partial pressure of LPG in a gaseous phase. Accordingly, it is difficult to dissolve in the culturing liquor the necessary amount of LPG for culturing the microorganisms in advance.

Referring to Figure 2.1, the apparatus comprises a fermentation liquor **1** which is stirred by a motor **2**. LPG is mixed with air through a flow rate control valve **3** and by a mixer **4**, and blown into the culturing liquor from an aeration pipe **9** through an automatic on-off valve **7** and a check valve **8**.

The vent gas which comprises the nonutilized gas leaves the top of the tank and enters a mist separator **10** where the mist is separated and discharged into the atmosphere through an automatic on-off

FIGURE 2.1: APPARATUS FOR ADDING LIQUEFIED HYDROCARBON GAS TO A CULTURE MEDIUM

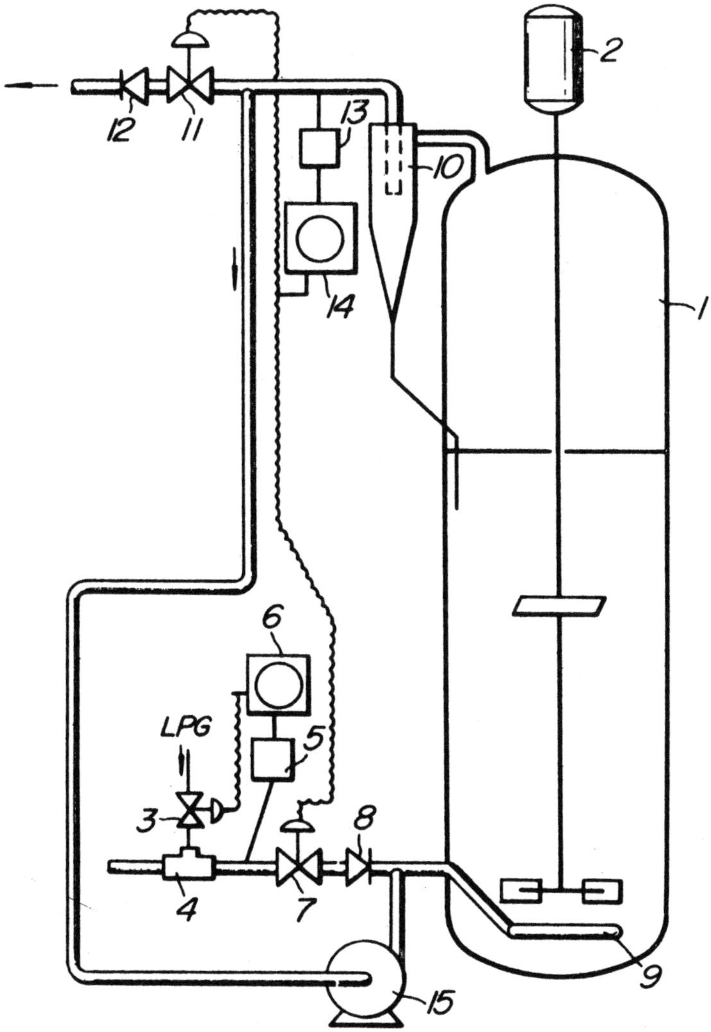

Source: O. Imada, K. Hoshiai and M. Tanaka; U.S. Patent 3,645,846; February 29, 1972

valve **11** and a check valve **12**. The LPG concentration in the air blown to the bottom of the tank is controlled by actuating the flow rate control valve **3** by means of a LPG concentration detector **5** and a LPG concentration controller **6**.

The LPG concentration in the vent gas is detected by a LPG concentration detector **13** and when the LPG concentration exceeds a predetermined value, an automatic on-off valve controller **14** transmits an "off" signal to the automatic on-off valves **7** and **11**, whereby the air supply system is closed.

On the other hand, the vent gas can be recycled to the aeration pipe **9** by a blower **15** and reblown into the tank. Accordingly, as the vent gas is recycled, the LPG concentration is lowered and the concentration is detected by the LPG concentration detector **13**.

When the LPG concentration becomes less than the predetermined value, the automatic on-off valve controller **14** transmits an "on" signal to the automatic on-off valves **7** and **11**, thus opening the air supply system, whereby a mixture of LPG and air is blown into the tank and the vent gas is discharged from the system. The foregoing operation is automatically repeated during the culturing period by the apparatus of this process, and thus the fermentation proceeds smoothly.

Example: In a fermentation tank which is 2 m. in diameter and 6 m. high, is placed 10 $m.^3$ of an aqueous fermentation medium containing 0.05% of KH_2PO_4, 0.05% of $MgSO_4 \cdot 7H_2O$, 0.01% of $MnSO_4 \cdot 4H_2O$, 0.01% of $FeSO_4 \cdot 7H_2O$, 0.30% of $(NH_4)_2SO_4$, 0.5% of yeast extract and 10 mg./l. of thiamine.

The solution is heated to 110°C. for 5 minutes to achieve sterilization and then cooled to 30°C. 300 l. of a precultured liquor of *Candida rigida* No. 113 (butane-assimilating microorganism) is inoculated into the sterilized fermentation solution.

The process is begun by stirring at a rate of 250 rpm and at a temperature of 30°C. The culturing solution is maintained at a pH of 6.0 during culturing by adding NH_3 gas thereto and by means of a pH controller. Culturing is conducted by supplying n-butane to the fermentation medium at a butane-gas concentration in the gas mixture of about 1.8% by automatically opening or closing the vent and gas supply valves so that the butane gas concentration in the vent gas can be controlled to about 0.05% or less. The flow rate of the recycling blower is adjusted to 30 Nm.^{3}/min., and the pressure difference between the inlet and the outlet of the blower should be from 0.4 to 0.5 kg./cm.^{2}.

After about 40 hours of culturing, the propagation of the cells of microorganisms reaches a maximum and a culture liquor containing 27 g./l. of cells of microorganisms is obtained. The culture liquor is then subjected to a centrifugal separation, and after drying the cells of microorganisms, 280 kg. of the dry cells are obtained. In this culture, the total amount of butane supplied is 1,080 kg., and the assimilation efficiency is 27.8%.

On the other hand, when culturing is conducted under the same conditions except that recycle is not utilized and the butane gas concentration in the aeration is adjusted to about 0.8 to 1.8%, the propagation of the cells of microorganisms did not reach 10 g./l. even after 40 hours of culturing.

The total butane supplied is 2,450 kg. and the assimilation efficiency is about 4%. In the latter case, the amount of butane dissipated into the vent gas is 2,040 kg. Thus about 83% of the total butane supplied is discharged out of the culturing system and not utilized.

SOLID HYDROCARBONS

Hydrocarbons can be fermented by microorganisms with the production of hydrocarbon soluble products and/or water-soluble products. Normally solid hydrocarbons are often preferred in fermentation processes. However, those hydrocarbons which are solids at fermentation temperatures and which can be utilized by microorganisms in fermentation processes are difficult to disperse in the fermentation system.

As a result, fermentation of solid hydrocarbons is a slow process as the hydrocarbons are not in a form readily available for attack by the microorganisms. An additional disadvantage of utilizing solid hydrocarbons in fermentation processes is that the solid hydrocarbons cannot be readily pumped or otherwise transmitted to or through the fermentation system.

G.H. Wegner; U.S. Patent 3,337,413; August 22, 1967; assigned to Phillips Petroleum Company has described a fermentation process wherein a normally solid hydrocarbon is solubilized in a liquid hydrocarbon of different chemical structure and the liquid hydrocarbon containing the normally solid hydrocarbon is employed in a hydrocarbon fermentation process.

In the hydrocarbon fermentation process hydrocarbon utilizing microorganisms are placed in an aqueous mineral salts medium. Hydrocarbon utilizing microorganisms include yeasts, molds and bacteria. Suitable bacteria include Eubacteriales, Actinomycetales, Chlamydobacteriales, Myxobacteriales, and Spirochaetales. The microorganisms are cultured in various mixtures of mineral salts dissolved in distilled water and with a normally solid hydrocarbon as a carbon source.

The hydrocarbons employed in the fermentation process are those hydrocarbons normally solid at the fermentation temperature and capable of being utilized by microorganisms in fermentation processes. These normally solid hydrocarbons include straight, branched and cyclic saturated and unsaturated hydrocarbons. The saturated straight chain hydrocarbons having at least 20 carbon atoms per molecule are preferred.

The normally solid hydrocarbon is solubilized by placing the solid hydrocarbon in a liquid hydrocarbon solvent of different chemical structure, thereby providing a means whereby the solid hydrocarbon can be readily handled, emulsified and fermented.

It is normally preferred that the liquid hydrocarbon solvent be of such a molecular structure as to be relatively resistant to attack by the particular species of microorganisms utilized. In this way, the liquid hydrocarbon solvent acts as an inert carrier for the solid hydrocarbon undergoing fermentation.

In some circumstances, it may be permissable, or even desirable, that the liquid hydrocarbon component, as well as the normally solid hydrocarbon, undergo fermentation. Suitable liquid hydrocarbon solvents include aromatic hydrocarbons, and straight, branched and cyclic saturated and unsaturated hydrocarbons liquid at the fermentation temperature.

In the hydrocarbon fermentation process, the selected hydrocarbon utilizing microorganisms are placed in a suitable aqueous medium. The liquid hydrocarbon solvent containing the normally solid hydrocarbon can then be placed on the surface of the aqueous medium, forming a hydrocarbon-aqueous medium interface, or circulated through the aqueous medium containing the hydrocarbon-utilizing microorganisms. The aqueous medium containing the microorganisms can be circulated through the liquid hydrocarbon solvent containing the normally solid hydrocarbon, or an emulsion of the liquid hydrocarbon solvent containing the normally solid hydrocarbon and the aqueous medium can be maintained throughout the fermentation process.

Example: In this example, runs were conducted to demonstrate the increased rate at which protein is produced by fermentation of eicosane dissolved in a liquid hydrocarbon as compared with the rate of protein production in the fermentation of eicosane as the solid.

The hydrocarbon liquid solvents employed in the fermentation runs were decane, a propane-dewaxed lubricating oil having a viscosity of 93.59 SUS at 100°F., a Cleveland open cup flash of 395°F. and an API gravity of 33.4, and Decalin. The fermentation runs were conducted with a pure yeast culture which grows well on C_{12} to C_{20} n-paraffinic hydrocarbons, but not on decane or Decalin. The culture employed in these experiments have been grown on eicosane. An aqueous mineral salts medium containing the following amounts of salts dissolved in each liter of solution was prepared.

Salt	Concentration, g./l.
KH_2PO_4	3
$(NH_4)_2HPO_4$	2
NH_4NO_3	1
$MgSO_4{\cdot}7H_2O$	0.4
$MnCl_2{\cdot}4H_2O$	0.002
$FeSO_4{\cdot}7H_2O$	0.02
$CaCl_2$	0.02
$CuSO_4{\cdot}5H_2O$	0.002
$ZnCl_2$	0.002

In each run, one ml. of the liquid hydrocarbon solvent was added to 50 ml. of the mineral salts medium in a 250 ml. flask. In those runs in which the normally solid hydrocarbon eicosane was dissolved in a hydrocarbon solvent, the 1 ml. of solution employed contained 50% by weight of eicosane and an equal weight of liquid solvent.

The mixture of liquid hydrocarbon and mineral salts medium was inoculated with 0.5 ml. of the yeast culture and then shaken at

200 rpm for 48 hours at 28° to 30°C. The extent of the yeast growth was then determined by protein analysis.

The quantity and results obtained for each of the runs from the fermentation of eicosane, decane, the propane-dewaxed lubricating oil, and certain mixtures thereof were as follows:

Run No.	Hydrocarbon	Protein in Culture, grams/liter
1	Decane	0.02
2	Eicosane	0.06
3	Dewaxed lube oil	0.26
4	Decane plus eicosane	1.00
5	Dewaxed lube oil plus eicosane	1.84

In the fermentation of a 1:1 mixture of Decalin and eicosane, shaken at 28° to 30°C., a heavy growth of yeast was observed after five days. When subjected to comparable conditions, Decalin in the absence of eicosane produced no yeast growth, and eicosane alone yielded only a poor growth of yeast.

The above results demonstrate the increased rate at which protein is produced by the fermentation of a normally solid hydrocarbon when the solid hydrocarbon is dissolved in the liquid hydrocarbon relatively resistant to attack by the microorganisms used.

Growth Promoting Techniques

USE OF CHEMICALS

Effect of Zinc on Yeast Growth and Protein

J.J. Cavallo and P.R. Hines; U.S. Patent 3,616,249; October 26, 1971; assigned to Atlantic Richfield Company have reported a process for improving the rate of growth and protein content of *Candida intermedia* grown on normal alkanes by providing a nutrient solution containing from about 8,000 to 10,000 micrograms of zinc, as a soluble zinc salt, per liter of nutrient.

The fermentation reaction is carried out in the presence of a nutrient medium containing a source of nitrogen, a source of phosphate, and various trace metals including zinc, preferably present as zinc sulfate. The zinc, however, may be present as any soluble nontoxic salt, such as zinc acetate, zinc acetylacetonate, zinc chloride, zinc citrate, zinc tartrate, etc. Similarly, the trace metals are perferably present as the sulfate or chloride and may be present in any form which is soluble and nontoxic to the organism present in the system.

Under these conditions, fermentation times approached or exceeded 30 hours. The optimum fermentation times are determined by following the rate of growth. Initially, the growth rate is comparatively low. Subsequently, the rate of growth will increase to a higher value, known as the "exponential growth phase" and subsequently the cellular density in the system will become constant as the growth rate diminishes. The yeast cells are normally harvested at or near the end of the exponential growth rate phase. Under such growth conditions, the crude protein yields were approximately 30% using the processes of the prior art.

By increasing the concentration of the zinc in the nutrient medium by approximately fiftyfold, as compared with the zinc level in the usual nutrient medium, the fermentation time may be decreased with a substantial increase in crude protein yield.

For example, satisfactory growth and protein production occurs when the nutrient medium contains 800 micrograms of $ZnSO_4 \cdot 7H_2O$ corresponding to approximately 183 micrograms of zinc per liter. In contrast, however, it has been found that a fiftyfold increase in zinc content, 4×10^4 micrograms of $ZnSO_4 \cdot 7H_2O$ (9,120 micrograms of zinc) per liter, in the nutrient medium significantly increases the rate of growth and, in addition, increases the protein content of the yeast produced.

Two growth media of the following compositions were prepared:

	Medium I	Medium II
KH_2PO_4	7.5 g.	7.5 g.
NH_4Cl	7.0 g.	7.0 g.
$CaCl_2 \cdot 2H_2O$	0.5 g.	0.5 g.
$MgSO_4 \cdot 7H_2O$	3.8 g.	3.8 g.
$CuSO_4 \cdot 5H_2O$	80 μg.	80 μg.
KI	200 μg.	200 μg.
$FeCl_3 \cdot 6H_2O$	400 μg.	400 μg.
$MnSO_4 \cdot H_2O$	800 μg.	800 μg.
$Na_2M_0O_4 \cdot H_2O$	400 μg.	400 μg.
$ZnSO_4 \cdot 7H_2O$	800 μg.	40,000 μg.
Distilled H_2O	1,000 ml.	1,000 ml.

After adjusting the pH of the media to 5.5, the solutions were sterilized at 248°F. at a pressure of 15 psig for 55 minutes. After cooling, one liter of Medium I was introduced into a sterile 5 l. fermenter. The system was then inoculated with 1 l. of 48 hour culture of *Candida intermedia* grown on the above medium and 2.8 ml. of an n-alkane mixture containing normal alkanes varying in chain length from C_{16} to C_{22} was introduced.

Oxygen at a rate of 0.1 vol./vol./min. was applied to the system and the liquid phase was agitated by a propeller type impeller rotating at 750 rpm. The pH was controlled at from pH 5.5 to pH 5.7 by hourly additions of NH_4OH in amounts varying from 1.1 to 3.5 ml. per hour to a total of 63.1 ml. The hydrocarbon substrate was added simultaneously with the NH_4OH in amounts ranging from 0.7 to

2.3 ml. per hour to a total of 44.6 ml. The fermentation was allowed to continue until the NH_4OH demand had stopped. The cells were then harvested and submitted for total nitrogen analysis. The fermentation using Medium II was conducted in the same manner. The data obtained from the two experiments and the results of the nitrogen analysis are shown in the following table.

From the data shown in the table in addition to a substantial decrease in fermentation time, there is a significant increase in crude protein yield. The fermentation time was decreased by 26% while the crude protein yield was increased by 26% using Medium II, with the high zinc content, as compared with Medium I.

Run No.	$ZnSO_4$ conc., μg/l	Fermentation time	Cell yield	Total nitrogen	Protein yield
1	800	27	82.3%	6.00%	30.8%
2	4×10^4	20	81.3%	7.58%	38.5%

While these data illustrate the unexpected increase in protein yield even at lower fermentation times for C_{16} to C_{22} n-alkanes, similar results would be obtained using other n-alkanes. It is intended that n-alkanes generally and n-alkanes having a chain length of from about C_{12} to C_{22} in particular be encompassed within the scope of the process.

Experimental data available indicate the preferable zinc content in the nutrient medium to be about 9,000 micrograms per liter but advantageous results may be obtained using zinc contents ranging from about 8,000 to 10,000 micrograms per liter of nutrient medium.

Addition of Alcohol and Fermentation Waste Liquor

I. Takeda, T. Iguchi, K. Tsuzuki and T. Nakano; U.S. Patent 3,674,640; July 4, 1972; assigned to Asahi Kasei Kogyo K.K., Japan have shown that yeast can be formed by continuously feeding a medium consisting of hydrocarbon, inorganic nitrogen, inorganic salts and organic source of nitrogen to which the fermentation waste liquor previously used and a small amount of an alcohol are added into an emulsification tank, mainly effecting formation of the hydrocarbon emulsion in the first tank as described below.

This method provides for the cultivation of hydrocarbon-consuming yeasts by a continuous fermentation process which comprises

(a) Cultivating *Torulopsis petrophilum, Candida petrophilium* or *Brettanomyces petrophilum,* a species of yeast capable of consuming hydrocarbons as the carbon source in a medium containing a hydrocarbon fraction boiling at temperatures in a range of 200° to 360°C.;

(b) Aerobically conducting the cultivation in an apparatus consisting of an emulsification tank mainly for forming emulsion of the hydrocarbon and an aqueous solution and a main fermentation tank for effecting cell formation or fermentative production of useful substances, the tanks being connected in series;

(c) Using as the medium in the emulsification tank a medium containing the hydrocarbon at a high concentration, the fermentation waste liquor from the main fermentation tank and an alcohol at a concentration of 0.5% by volume or lower based on the entire volume of the medium; and

(d) Feeding into the main fermentation tank the emulsified fermentation liquor obtained in the emulsification tank.

As the alcohol used in this method may be used any lower alcohols such as methanol and ethanol as well as any higher alcohols such as lauryl and cetyl alcohols. The amount of alcohol to be added may be varied and preferably is in a range from 0.03 to 0.5% by volume of the entire medium.

Effects of the addition of the fermentation waste liquor and alcohol in a small amount are described in detail by the experiments given below.

Experiment 1: Experiment was made using *Candida petrophilum* ATCC No. 20226. Composition of the medium in the first tank is as follows.

Straight chain paraffins containing from 14 to 17 carbon atoms (purity, 98% or higher): 15% (by vol./vol.)
Manganese sulfate: 0.004% (by weight/volume)
Magnesium sulfate: 0.05% (by weight/volume)
Calcium chloride: 0.01% (by weight/volume)

Urea: 0.3% (by weight/volume)
Ammonium sulfate: 0.2% (by weight/volume)
Ferrous sulfate: 0.005% (by weight/volume)
Corn steep liquor: 0.1% (by weight/volume)

The medium of the above composition, pH 4.5, 4 l., is placed in the first tank, 8 l. in capacity, and maintained at the same concentrations of the nutrient sources. In fact, the consumed amounts of the inorganic salts, nitrogen source and organic nutritive source as determined by the yield of the product are constantly supplemented. In the use of fermentation waste liquor for emulsion formation, the shortage found in consideration of the composition of the waste liquor is supplied prior to the operation.

Aerobic cultivation was carried out in the second tank. That is, 20 l. in volume of the medium was placed into 40 l. in volume of the tank, in which no hydrocarbon was added and an aqueous solution of the same composition as above except for the hydrocarbon was continuously fed.

Table 1 shows the results of determination of the time of emulsion of formation (residence time) in the first tank and the efficiency of cell formation in the second tank, in the case where the fermentation waste liquor is added at a ratio from 5 to 75% by volume on total medium. There are observed reduction in time required for emulsifying the hydrocarbon by the addition of the fermentation waste liquor in the first tank and rise in ratio of the formation.

TABLE 1: EFFECTS OF THE FERMENTATION WASTE LIQUOR UPON THE EMULSIFICATION AND MYCELIAL FORMATION

Concentration of the fermentation waste liquor, percent (by volume)	Time of emulsification (1st tank), hr.	Dried mycelia (g.)./hydrocarbon added (g.), percent
5	5	92
10	5	95
25	4	99
50	3.5	103
75	3	103
0	7	91

Fermentation conditions in the first tank; 30°C., 500 rpm, aeration 60%, liquor volume 4 l., adjusted to pH 4.5 to 5.0 by aqueous ammonia. Fermentation condition in the second tank: 30°C., 400 rpm, aeration 100%, residence time 4 hours.

Experiment 2: The same strain and medium were used as in Experiment 1 except for the fermentation waste liquor which was added in the first tank at a concentration of 25% by volume on total medium and examined for the effect of addition of the alcohol in the first tank. Table 2 shows the results.

TABLE 2: EFFECTS OF THE ALCOHOL ADDITION (PERCENT)

Concentration of alcohol by volume	0.05%	0.1%
Alcohol:		
Methanol	106	101
Ethanol	108	118
Propanol	111	118
Isopropanol	112	119
Butanol	110	117
Pentanol	109	116
Hexanol	106	106
Heptanol	110	108
Capryl alcohol	111	113
Nonyl alcohol	113	111
Decyl alcohol	117	110
Lauryl alcohol	108	111
Myristyl alcohol	105	112
Cetyl alcohol	110	112
Octadecanol	108	117
Control	99	

The figures in the table indicate; dried cells in gram per added hydrocarbon in gram x 100. As clearly seen from Table 2, simultaneous addition of the fermentation waste liquor and alcohol produces remarkable synergistic effects.

Example: Torulopsis petrophilum ATCC No. 20225 was cultivated by the process. The alcohol used was propanol. Propanol was continuously added to the first tank at a concentration of about 0.05% by volume. Continuous fermentation was conducted using as the second tank a Waldhof type tank. The conditions of fermentation and treatment after separation of the cells resulted in the dried mycelia in a yield of 104 g./100 g. of the hydrocarbon added. Efficacy of the cells separation by the use of a Westfalia separator was 92%. The cells thus obtained can be utilized as they are as a protein source in the animal feed.

Unsaturated Organic Acids

In a fermentation process using a hydrocarbon as the source of carbon, *J.F. Herndon, M.B. Perkins, N.J. Stevens and J.D. Douros, Jr.; U.S. Patent 3,508,927; April 28, 1970; assigned to Esso Research and Engineering Company* add growth stimulants to the fermentation zone, the growth stimulants being selected from the class of unsaturated organic compounds consisting of acrylic acid, fumaric acid and oleic acid. The process contemplates use of any catalytic compound containing one or more

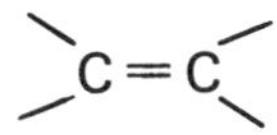

groups which imparts a characteristic growth promoting action.

The olefins which suitably impart such growth promoting characteristic are monoolefins of the type:

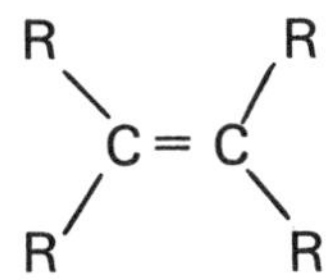

and diolefins of the type:

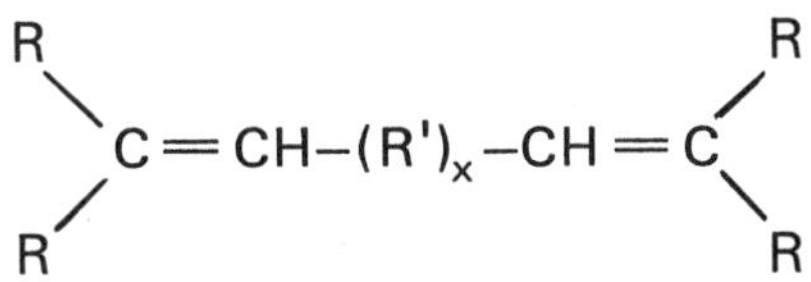

wherein R may be hydrogen or any aliphatic hydrocarbon group containing from about 1 to 10 carbon atoms wherein R' may be any aliphatic hydrocarbon containing from about 1 to 10 carbon atoms, and wherein x represents an integer of from 0 to about 25.

The preferred olefinic hydrocarbons contain from about 11 to 30 carbon atoms per molecule. Nonlimiting examples of typical compounds are propylene, butene-1, isobutylene, hexene, heptene, octene, nonene, decene, hexadecene, octadecene, eicosene, butadiene, piperylene, isoprene, 1,5-hexadiene, 1,4-octadiene, 2,2-dimethylbutadiene, and the like.

Illustrative of the carboxylic acids suitable for use in the bisosynthesis process of this process are the acrylic acids, i.e., acids of the type:

$$C_nH_{2n-1}COOH$$

the fumaric acids, i.e., acids of the type:

$$C_nH_{2n-2}(COOH)_2$$

wherein n represents an integer of from 2 up to about 22. The preferred unsaturated carboxylic acids contain from about 8 to 22 carbon atoms per molecule. Nonlimiting examples of such acids include acrylic acid, the butenic acids, i.e., crotonic, isocrotonic, vinylacetic and methylacrylic acid, the pentenic acids, i.e., tiglic, angelic and senecioic acid, the hexenic acids, teracrylic acid, oleic acid, fumaric acid, maleic acid, itaconic, the xeronic acids, and the like.

Example 1: Arthrobacter ureafaciens were grown on an essentially C_{16} n-paraffin hydrocarbon containing about 10 ppm aromatics. The composition of the inorganic salts solution utilized was as follows:

Mineral Salt	Feed Level, Weight
$CaCl_2$	0.025
$FeSO_4 \cdot 7H_2O$	0.004
KCl	0.10
$MgSO_4$	0.025
Na_2SO_4	0.05
85% H_3PO_4	0.25
Water	99.5
Total	99.954

The growth conditions established prior to feeding the slurry to the continuous reactor were a 4 l./min. air rate; a 20 g./hr. C_{16} hydrocarbon feed rate; a 2 l./hr. mineral salts feed rate; 13 N ammonium hydroxide fed on demand to control pH; a temperature of 30° to 37°C. and agitation at 350 to 1,500 rpm.

About 2 l./hr. of approximately 9.5% by weight solids aqueous slurry produced as outlined above was fed to a vessel suitable for continuous reaction having a volume of 7.5 l. in which *Arthrobacter ureafaciens* was growing at steady state. About 0.2 g./hr., i.e., 0.01 wt. percent, of oleic acid was added to the aqueous slurry feed. Sufficient air was then bubbled through the aqueous slurry. The experiment was conducted at temperatures of 30° to 37°C. The pH was maintained at a pH of approximately 7.0 throughout the run. The air rate was 22 cc/min./cm.^{2} of reactor cross-section and the agitation was effected at 800 rpm.

At a steady state, representative samples were taken from the reactor vessel. These samples analyzed for bacteria cell and hydrocarbon concentration. In order to demonstrate the feasibility to yield bacteria concentrations substantially exceeding those obtainable in conventional biosynthesis baths (where no growth promoting agent was utilized), a biosynthesis was accomplished essentially identical to the foregoing save for the absence of oleic acid. The results are shown below for comparative purposes:

Additive	Cell Concentration, g./l.
Oleic acid	14.4±1
None	12.6±1

Example 2: The following example was effected to evaluate growth promoters in a continuous operation. A run in which *Arthrobacter ureafaciens* had been grown continuously in a manner similar to that of Example 1 and for a duration greater than 300 hours was completed. In this run, oleic acid and 1-hexadecene were studied at various feed levels. The effect of growth promoters with a 1 weight percent total hydrocarbon feed is shown in the table below. The additive is given in weight percent feed concentration and cell growth rate in g./hr.

Additive	Rate
0.1% oleic acid	19.0±1
0.01, 0.02, 0.03% oleic acid	16.4–19.5
0.1, 0.01% 1-hexadecene	16.6±1
After additive removal	12.6±1

From the foregoing data it can be concluded that oleic acid and 1-hexadecene are effective as growth promoters. Additionally, the data show that 0.1 weight percent oleic acid evidences slightly better results than 0.01 to 0.03 weight percent oleic acid.

Example 3: In a manner similar to that of the preceding examples, azelaic acid was employed as a growth promoter and was compared with a run employing no promoter.

Comparative data showed that *Arthrobacter ureafaciens* when grown on C_{16} hydrocarbon feed stock with 50 ppm of azelaic acid added to the medium stimulated growth and hydrocarbon utilization of the bacteria in a 48 hour period by 24% over that of the control. It also increased the amount of hydrocarbon utilized and thus allowed more carbon to go to chemicals. It is noteworthy that the concentration of azelaic acid required was only 50 ppm and, advantageously, this concentration would not influence the total cost of product produced.

Nonionic Surface Active Agent

R.I. Leavitt and I.J. Heilweil; U.S. Patent 3,293,145; December 20, 1966; assigned to Mobil Oil Corporation have described a process for incubating a microorganism with an aqueous mineral salts medium in the presence of oxygen and, as sole source of carbon, an aliphatic hydrocarbon and, further, in the presence of a small amount of a nonionic surface active agent which acts as a growth stimulator.

The preferred microorganism is an Achromobacter, such as *A. xerosis, A. gutatus, A. superficialis, A. parvulus,* and *A. cycloclastes.* Also useful is the genus Nocardia, particularly *N. salmonicolor,* but also including *N. asteroides, N. opaca, N. corallina,* and *N. rubra.*

The aliphatic hydrocarbon which forms the sole source of carbon for the microorganism is a saturated or unsaturated, straight or branched chain hydrocarbon having up to 20 to 30 or more carbon atoms. Saturated straight chain alkanes having up to 20 carbons are preferred, particularly those which are liquid at the temperatures employed.

The amount of hydrocarbon is that required to provide carbon for growth and energy and may be chosen for any desired growth period, although it may also be added from time to time to a culture mixture as may be necessary. It is desirable to add only the required amount so as to avoid separation problems at the end of the growth period.

The nonionic surface active agent is preferably a compound having an aromatic nucleus, such as a phenyl ring, substituted by a side chain of hydrophilic character such as a polyoxyethylene group, and by a side chain of lipophilic character such as an alkyl group. Agents of this kind are frequently referred to as polyoxyethylene alkyl aryl ethers, obtainable as by reacting an alkylphenol with ethylene oxide. The agent is useful in concentrations as low as 0.001% by weight of the aqueous mixture which is incubated, but usually is 0.01 to 0.1% of the mixture and may range up to 0.5, 1, or 5% or more.

Generally, the lower concentrations are preferred. It will be noted that the agent is a nonionizing compound. Ionizable agents are unsuitable, it having been found that they do not stimulate cell growth.

Example 1: A soil-isolated organism identified as a member of the genus Achromobacter was inoculated into several 250 ml. Erlenmeyer flasks, each containing 50.0 ml. of the mineral salts medium whose qualitative and quantitative composition is listed above and to which 0.3 ml. decane had been added. The decane functioned as the sole carbon source. Various additions which were made to the flasks were incubated at 25°C. with shaking and the optical density of the various cultures determined at periodic intervals. Data obtained are as follows:

Run	Additive	Final Conc. Percent	Agent Conc., Percent	Optical Density At 24 Hrs.	At 48 Hrs.	At 72 Hrs.
1	–	–	–	0.05	0.24	0.88
2	Renex 688*	0.01	0.01	0.19	4.72	4.16
3	Renex 688	0.001	0.001	0.15	0.66	1.44

*A nonionic surface active agent.

After 38 hours decane could still be detected in flasks 1 and 3 but the decane in flash 2 was exhausted. Additional decane (0.5 ml.) was added to each flask at this time and the flasks reincubated an additional 24 hours. As may be seen, the addition of nonionic agent resulted in a significant increase in the growth rate of the culture.

Thus, after 48 hours of incubation, flask 2 exhibited an optical density of 4.72 as against a value of 0.24 for the control, or an increase of 1,870%; after 72 hours, the increase was 370%. When 0.001% of agent was used, the respective increases over the control were 175% and 63%. It may be added that additional controls were run lacking either cells or the decane, but no growth was detected in these. In an experiment using an anionic surface active agent in place of the Renex 688, the experiment being otherwise the same, there was no stimulation of cell growth. And in a further experiment using a cationic surface active agent there was no stimulation of cell growth.

The data are expressed above as optical density, and the relation between optical density and cell growth is as follows: an optical density of 1.0 is equivalent to approximately 2.1 g./l. of cells, dry weight.

Example 2: Cell growth was carried out on a larger scale using a bacterium identified as *Nocardia salmonicolor* 107-332. It was grown in two 5 l. fermenter vessels, each containing 3 l. of mineral salts medium as used in Example 1. One of the vessels contained Renex 688 at a concentration of 0.001% and the other vessel contained no agent, serving as a control. During the entire incubation period n-butane and air were bubbled through the culture liquid in a ratio of 10 parts air to one part butane. The vessels were incubated with agitation at 30°C. and growth in each fermenter determined in terms of optical density increase at regular intervals. Both cultures grew at approximately the same rate during the first nine days of incubation.

However, after this period, the culture without added agent stopped growing and exhibited a decrease in density; it therefore was harvested. At this point, the cells of the first culture continued to grow and divide but at a lower rate than before; they continued to increase in density for a total of 21 days at which time they too were harvested. The untreated cells produced a yield of 12.8 g./l.; the treated cells 18 g./l. A comparison of the dry weights of the two cultures showed that the addition of the small amount of agent resulted in an increase in cell yield of approximately 40%.

Cell composition was analyzed periodically in both cultures but no differences were found in the protein, lipid, or polymer [poly(beta-hydroxybutyrate)] contents from the beginning to the termination of the experiment.

In another run, which was just like the foregoing control run, 0.001% of the surface active agent was added to the culture at the time it reached its maximum phase of growth, i.e., at the point at which growth ceased; and a noticeable stimulation of growth amounting to 10 to 15% took place.

Increasing the growth rate of microorganisms is of importance in view of the ability of many of them to synthesize useful products such as proteins, amino acids, vitamins, lipids, polymers, and other compounds of value. These products may be formed in the cells or secreted by them into the medium in which the microbe is growing, and in either case, are recoverable. Their rate of formation increases with the rate of cell growth, and as a result, less time is required before harvesting of the cells may take place.

In some cases, the process provides greater yields of cells than are otherwise attainable and is also of importance to the microbial oxidation of hydrocarbons. Thus, certain hydrocarbons, such as n-decane, are not only capable of being utilized by a microorganism as the carbon source therefor but also, and during the course of being utilized, are converted by the microorganism into various oxygenated hydrocarbon products, including aldehydes, ketones, acids, and esters, many of which are of higher value than the starting hydrocarbon. A frequent difficulty in microbial oxidations is the relatively slow rate of growth of the microbe. By aid of this process, this difficulty may be considerably lessened.

Use of Ozone

S.V. Chepigo, S.I. Belenky, V.S. Somov, V.G. Voronkov, M.A. Belavsky, O.G. Korotaev, G.S. Rodionova, N.S. Maximenko, B.A. Glazman, L.A. Belova and G.I. Vorobiova; U.S. Patent 3,654,084; April 4, 1972 have described a process for producing yeast, predominantly fodder type yeast, by continuous cultivation of yeast in an aqueous nutrient medium containing plant tissue hydroyzates or n-paraffins as the source of carbon and aerating the nutrient medium with air to which ozone is added in a concentration of 10^{-6} to 1.5 volume percent.

Methods are known for producing yeast on aqueous nutrient media containing as the source of carbon, hydrolyzates of plant tissues, where inhibitors of yeast growth, such as furfural, hydroxymethylfurfural, tannin, physiologically active colloids, etc., have an adverse effect on the processes of yeast respiration and propagation. This reduces the yeast yield and retards multiplication.

In order to increase the yeast yield, nutrient media must be greatly diluted (to about 1% concentration of reducing substances), which results in less efficient ultilization of the manufacturing facilities.

When employing known methods of growing yeast on aqueous nutrient media containing as the source of carbon, hydrocarbons, for instance, n-paraffins, the medium is aerated with air alone, or with air supplemented with oxygen. The above conditions, however, do not ensure sufficient intensity of growth, since hydrocarbons are oxidized with difficulty and utilized by microorganisms at low rates.

A method for growing yeast in an aqueous nutrient medium has been developed in which the medium is aerated with air to which ozone has been added. Ozone can be added in a concentration of 10^{-6} to 1.5 volume percent.

Ozone is preferably added to the air used for treating the aqueous nutrient medium when the medium is being prepared. Ozone can also be added to the air employed for aerating the aqueous nutrient medium in the process of yeast growing. Besides, ozone can be added to the air used for treating water prior to employing the water in preparing the nutrient medium.

The only carbon source used in preparing the aqueous nutrient medium is a substance selected from the group of substances consisting of hydrolyzates of plant tissues and hydrocarbons of the n-paraffin series.

Hydrocarbons employed comprise the paraffin fraction of petroleum consisting of compounds with 11 to 30 carbon atoms and boiling in the range of 240° to 360°C. Hydrolyzates of plant tissues employed comprise agricultural wastes, e.g., hydrolyzates of cotton seed hulls, corncobs, etc., wastes of the woodworking industry, and also wood hydrolyzates. Fermentation is carried out continuously in an aqueous medium having a standard composition of microelements and potassium, magnesium, phosphorus and nitrogen salts.

The pH of the nutrient medium is maintained in the range of 3 to 6, preferably in the range of 4 to 5.5. The optimum temperature of the fermentation process is usually in the range of 25° to 40°C. Ozone is one of the strongest oxidizing agents, and is capable of oxidizing diverse chemical compounds.

When treating hydrolyzates with ozone, the pH of the medium and the temperature and concentration of the ozone-air mixture is maintained at levels providing selective oxidation of inhibitors without oxidizing the carbohydrates essential for yeast nutrition.

Treating hydrocarbons of the n-paraffin series with ozone-air mixtures having an ozone concentration of 1.0 to 1.5 volume percent at a temperature of about 100°C. causes oxidation of the terminal groups of n-hydrocarbons and the oxidation of cyclic hydrocarbon impurities.

As a result, surface active organic acids are formed which stabilize emulsions of hydrocarbon nutrient media. The multifold increase in the phase interface on which yeast multiply promotes their propagation and increases the yield of biomass.

Ozone treatment of water before use for diluting hydrolyzates and preparing nutrient media from hydrocarbons of the n-paraffin series raises the redox potentials of the water to levels at which extraneous microorganisms and their spores, and also viruses perish completely.

Ozonized water is practically sterile. Moreover, ozonized water used immediately after preparation oxidizes inhibitors of yeast growth occurring in hydrolyzates and increases the stability of the emulsions of hydrocarbon nutrient media.

Example: Strain NP-4 of yeast *Candida guillermondii* was grown by the continuous method for 18 hours in a 1.5 l. fermenter (volume of substrate 0.5 l.) on an aqueous nutrient medium providing a standard content of microelements and salts of K, Mg, P and N, and containing 1 volume percent of purified paraffins, at a temperature of 32.0±0.5°C. and a pH of 4.2 to 4.5, with a 10 hour change of culture fluid. The culture fluid was aerated with air containing 10^{-4} volume percent of ozone. Yield of absolutely dry yeast 130 wt. percent on basis of paraffins consumed.

Use of Antibiotics

J.D. Douros, L.A. Naslund and C.J. McCoy; U.S. Patent 3,414,477; December 3, 1968; assigned to Esso Research and Engineering Co. have described a bacterial fermentation process for the preparation of a high protein-content product utilizing a C_1 to C_{35} hydrocarbon feed, where the improvement comprises increasing the concentration of essential amino acids in nutritionally balanced proportions by the incorporation of streptomycin in the hydrocarbon medium.

This process relates to a method for both increasing the microorganism growth rate and the hydrocarbon conversion rate of a cultivation process which comprises using varying concentrations of antibiotics in order to change the amino acid profile of a microorganism in order to make it a better and more balanced protein.

The biosynthetic process is applicable to the biosynthesis of all microorganisms, including bacteria and yeasts, capable of growth on C_{10} to C_{30} hydrocarbon feeds.

While this process is applicable to a broad scope of operable microorganisms, there are nine microorganisms which are especially suitable for hydrocarbon assimilation.

These microorganisms are tabulated hereinbelow along with with their corresponding ATCC registration numbers, which were secured by depositing samples with the American Type Culture Collection in Washington, D.C.

Microorganism Name	ATCC No.
Micrococcus cerificans	14987
Pseudomonas ligustri	15522
Pseudomonas pseudomallei	15523
Pseudomonas orvilla	15524
Alcaligenes sp.	15525
Cellumonas galba	15526
Brevibacterium insectiphilium	15528
Corynebacterium sp.	15529
Corynebacterium pourometabolum	15530

The process is particularly concerned with a technique for biologically synthesizing food protein by cultivating microorganisms on petroleum substrates. This latter type of protein synthesis is usually conducted in an aqueous biosynthesis bath containing a hydrocarbon feed, an inoculant of the microorganism to be grown, an aqueous growth medium, oxygen and other indispensable nutrients. This technique allows the use of hydrocarbon feeds, which are less expensive than carbohydrates, and does not usually require expensive growth factors such as vitamins, amino acids, etc. in order to insure proper microorganism cell growth.

Hydrocarbon feeds which can be utilized for this process are C_{10} to C_{30} petroleum hydrocarbon feeds, preferably gas oils boiling in the range of between 190° and 400°C., preferably between 190° and 320°C. Other suitable feeds are C_{10} to C_{30} normal and isoparaffins, cycloparaffins, monoolefins, diolefins, aromatics and mixtures thereof. A preferred feed is one which contains a substantial weight percentage, e.g., 70+ weight percent, of normal (straight chain) paraffin hydrocarbons having from 10 to 30 carbon atoms.

While the presence of branched, nonaromatic hydrocarbons in amounts of up to 30% by weight in the hydrocarbon feed can be tolerated, concentrations in excess of 10 weight percent of nonnormal, nonaromatic hydrocarbons are usually avoided since the preferred microorganism cells employed in this process are selective preferentially to normal hydrocarbons, especially small n-paraffins. Therefore, the use of branched paraffins is preferably avoided.

A most preferred hydrocarbon feed is a C_{10} to C_{30} feedstock which has been purified to reduce the level of aromatics, both polycyclic and monocyclic, to below 0.5 weight percent, preferably below 0.1 weight percent, more preferably below about 100 ppm.

Unexpected, desirable results are secured by the use of an antibiotic in the fermentation process in that the protein produced is a much higher quality and much higher in the essential amino acids.

Example: Arthrobacter species was treated in the presence of 250 gamma/100 ml. of media of streptomycin and streptomycin plus trichloroacetic acid treated Pharmamedia under standard conditions. This concentration of antibiotic had no effect on the growth of Arthrobacter species as shown in Table 1.

TABLE 1

	Arthrobacter Species		TCA Treated Pharmamedia [1,2]		Strepto-mycin +Pharma-media	Strepto-mycin
	Control	+Pharma-media (0.01%)[3]	+Strepto-mycin Solids	+Strepto-mycin Liquids		
g./l. Cells	11.33	11.82	11.04	12.05	11.96	11.54

[1] Traders Protein Div., Fort Worth, Tex., U.S.A.
[2] Trichloro acetic acid treatment was used to precipitate the protein from Pharmamedia. This was done to determine if the small amount of precipitated protein could be used to influence the metabolism of the organism, i.e. increasing its essential amino acid pools.
[3] 0.01 gms. of Pharmamedia was treated with 1.0 cc. of 10% trichloro acetic acid solution. The precipitated protein is designated as liquid and the insoluble fraction of the Pharmamedia is designated as solids.

TABLE 2

N/16 Basis	Control	Pharma-media	TCA Treated Pharmamedia		Strepto-mycin +Pharma-media	Strepto-mycin
			Strepto-mycin +Solids	Strepto-mycin +Liquid		
Percent S. Amino Acids	1.84	1.99	2.36	2.22	2.87	2.96
Percent Lysine	4.6	4.8	4.8	5.3	5.6	6.4

The process is concerned with the use of an antibiotic such as streptomycin in a fermentation process. Other suitable antibiotics are, for example, neomycin, chloromycetin, Vancomycin, Colymycin, penicillin.

The concentration of the antibiotic used may vary appreciably depending upon operating conditions. The concentration may range from 10 to 6,000 gamma/100 ml. of media. A preferred range is from about 100 to 500 gamma/100 ml. of media, such as about 250 gamma/100 ml. of media. The time period likewise may vary appreciably as, for example, from 24 to 72 hours as, for example, about 42 hours. The concentration of the hydrocarbon in the media is from 0.1 to 10% by volume, preferably in the range of from 1 to 2% by volume.

USE OF DISPERSIONS

Effect of Finely Divided Adsorbant

In the conversion of hydrocarbons to various products by the action of microorganisms, reactions have been notoriously slow and replete with undesired secondary reactions leading to loss of the desired products and the production of undesirable products. Such reactions, even on a small scale, have been matters of hours and even days.

The procedure of *R.G. Harris; U.S. Patent 2,769,750; November 6, 1956; assigned to Texaco Development Corporation* is concerned with the improvement in and acceleration of processes involving interactions between microorganisms and organic substances by effecting reaction in the presence of finely divided and relatively inert adsorbent material. The process discloses that the metabolism of hydrocarbons and oxygenated hydrocarbons by microorganisms is significantly catalyzed by the presence of a finely divided adsorbent. The contact of organic compound, adsorbent and microorganisms is effected in the presence of a nutrient particularly suited for the microorganism or microorganisms.

Organic materials whose utilization by microorganisms is accelerated by the presence of a finely divided adsorbent are broadly defined as hydrocarbons and oxygenated hydrocarbons comprising alcohols, aldehydes, ketones, esters and acids.

The process is particularly useful in the processing of hydrocarbons and oxygenated hydrocarbons containing more than 12 carbon atoms which are difficult to disperse in the nutrient-microorganism mixture and which are utilized only very slowly by microorganisms in the absence of the finely divided inert adsorbent. The processing of hydrocarbon oil fractions boiling above 400°F. such as gas oil, lube oil fractions and residual fractions is enhanced by the addition of a finely divided adsorbent.

The addition of a finely divided adsorbent increases the action of microorganisms on solids such as petroleum wax, petroleum resins, asphalt, solid alcohols and solid esters such as beeswax and Chinese wax and on high boiling liquid hydrocarbon fractions such as gas oil, lube oil and residual fractions. Although the catalytic action of a finely divided adsorbent on high boiling hydrocarbons and oxygenated hydrocarbons is noteworthy, the processing of lighter hydrocarbon fractions and oxygenated hydrocarbons is also improved by the presence of an adsorbent.

A naturally occurring earth or clay is the preferred inert adsorbent. Kaolinic and montmorillonitic clays have been found very satisfactory. Fuller's earth and Ione clay are particularly preferred natural clays.

Particle size distribution of the adsorbent and the method by which the substance to be treated is mixed therewith are important in increasing the efficiency of the process. Best results are obtained when the adsorbent is formed of particles 75 to 90% of which have diameters of less than 1.0 micron. The following example is cited to illustrate a preferred particle size distribution for a good adsorbent:

Particle Diameter in Microns	Percent Distribution in Samples
Less than 0.1	5
0.1 to 1.0	80
1.0 to 10.0	10
10 and above	5

The solid adsorbent is premixed with the hydrocarbon or oxygenated hydrocarbon substrate. The total amount of adsorbent added is in the range of 1 to 10 volumes of adsorbent per volume of hydrocarbon or oxygenated hydrocarbon being treated. The proportions of adsorbent to hydrocarbon depend on the nature of the substance to be treated; the polarity and ease of adsorption of the treated material are important in determining how much adsorbent is to be used. Normally, 1 to 3 parts of adsorbent to 1 part of liquid or solid hydrocarbon are employed.

Sufficient substantially anhydrous adsorbent is mixed with a substance to insure a thoroughly homogeneous mixture. It is preferred to add sufficient adsorbent to convert the hydrocarbon-adsorbent mixture to a crumbly, powdery phase. This dry mixture is termed an adsorbent complex. In this modification, the adsorbent complex is introduced into a nutrient solution that has been inoculated with the desired microorganism.

The addition of the finely divided adsorbent exerts a catalytic action on the rate at which a wide variety of microorganisms utilize hydrocarbons and oxygenated hydrocarbons. In general, the action of all hydrocarbon-consuming microorganisms in processing hydrocarbons and oxygenated hydrocarbons is accelerated by the presence of a finely divided adsorbent. Hydrocarbon-utilizing bacteria, actinomyces, molds, yeasts and fungi may be employed in the process.

Nutrient solution is employed in the reaction mixture in an amount such that the hydrocarbon or oxygenated hydrocarbon substrate is present in an amount approximately 1 to 10% by weight of the entire reaction mixture including adsorbent and nutrient solution. The portion of substrate present in the total reaction mixture is in the lower portion of the 1 to 10 percent range.

This process is illustrated in the following example which demonstrates the catalytic effect of a finely divided adsorbent on the conversion of a large quantity of paraffin wax to degradation products. 100 g. of paraffin wax was mixed with 200 cc of nutrient solution and contacted with a wax-destroying species of Pseudomonas; after three months, the wax sample was found to have undergone no substantial decomposition.

Another 100 g. sample of the same paraffin wax stock was mixed with 150 g. of adsorbent clay, and 200 cc of nutrient solution; this mixture of wax, adsorbent and nutrient was inoculated with the same wax-destroying Pseudomonas species with the result that the wax was entirely converted within 72 hours to oxygen-containing decomposition products.

The catalytic effect of a finely divided adsorbent on the utilization of hydrocarbons by hydrocarbon-consuming microorganisms has also been demonstrated on a microscale employing small quantities of hydrocarbons. In this test, the results of which are shown in the accompanying table, the increased conversion of high molecular weight hydrocarbons by microorganisms in the presence of a finely divided adsorbent is shown by the increased oxygen utilization obtained when the hydrocarbon is subjected to microbial decomposition in the presence of a finely divided adsorbent.

In this test, homogeneous powdery hydrocarbon-clay complexes were prepared by heating approximately 100 g. of the hydrocarbon substrate with 150 g. of clay at a temperature of approximately 120°C. Sufficient of this complex was then added to 50 ml. of nutrient solution in an 8 ounce bottle to make a 2.5% hydrocarbon concentration in the total culture medium. A parallel series of experiments was set up to demonstrate the normal rate of hydrocarbon conversion by microorganisms in the absence of the catalytic adsorbent material; in this series, sufficient hydrocarbon was added to 50 ml. of nutrient solution to impart to the resulting culture mixture a hydrocarbon concentration of 2.5%. Both series of reaction mixtures were then inoculated with a hydrocarbon-clay culture of the test microorganisms; each reactor bottle was seeded with 2.5% of the appropriate seed culture.

The air-filled reactor bottles were then placed in a shaking apparatus and were maintained during shaking at a temperature of 28° to 30°C. When growth was observable in the reactor bottles, shaking was stopped, nitrogen gas introduced to bring the pressure back to atmospheric and the volume of oxygen utilized calculated from the percentage of residual oxygen in each bottle. The following table clearly demonstrates that the presence of finely divided adsorbent effects a two to seven-fold increase in oxygen utilization micro evaluation of the process:

Organism	Hydrocarbon	Hours in Shaker	Ml. O_2 utilized	
			Hydrocarbon	Hydrocarbon clay complex
Achromobacter sp. TDC (B-148).	Paraffin wax	26	10.6	40.0
Alcaligenes sp. TDC (A-47).	do	22	14.5	36.1
Aspergillus sp. TDC (946).	do	154	4.0	29.4
Aspergillus sp. TDC (946).	Arabian crude oil.	154	37.2	55.6

The results demonstrate the catalytic effect exerted on the action of microorganisms by the use of a finely divided adsorbent. The presence of clay caused a four-fold increase in the utilization of paraffin wax by a species of Achromobacter, approximately a two and one-half-fold increase in the utilization of paraffin wax by a species of Alcaligenes, a seven fold increase in the utilization of paraffin wax by a species of Aspergillus and almost a two-fold increase in the utilization of Arabian crude oil by the same Aspergillus species.

Increasing Microbial Fermentation with Zeolites

The fermentation of hydrocarbons presents unusual problems because of the insolubility of hydrocarbons in water. Microbes usually act and thrive in aqueous media, and normally the application of microbes involves the placement and maintenance of the microbes in aqueous media which contain available food material. If a hydrocarbon is to be the only or the main source of food in the aqueous medium containing microbes, it must somehow be made available to the hydrophilic microbe.

R.L. Raymond; U.S. Patent 3,224,946; December 21, 1965; assigned to Socony Mobil Oil Company, Inc. has increased the rates of microbial conversions by adding to the medium that contains or is to contain the substrate, such as the hydrocarbon, and the microbial cells a zeolite in comminuted form, suspending it in the medium, effecting, prior or in situ, occlusion of the substrate thereon and then allowing the microbial cells to oxidize or otherwise attack the occluded substrate particles.

The substrate, as, for example, a hydrocarbon, is adsorbed on the molecular sieves or within the crystal dimensions so that it is broken down into a large number of small particles which are kept so by physical entrapment and adsorption. With reference to hydrocarbons it is observed that regardless of their volatility the hydrocarbon particles are held suspended in the water for long time periods in the absence of microbial attack, and in the presence of microorganisms, the hydrocarbons and the microbes in the mixture become and remain uniformly distributed throughout the medium, as is observed microscopically. The cells become attached to the sieve particles and grow rapidly on the occluded hydrocarbon particles. The processes thus comprise microbial action in the presence of zeolite carriers.

Example: To a 250 ml. bottle is added 50 ml. of water containing the following salts, the amount in parentheses indicating the number of parts used per 1,000 parts of water, ammonium sulfate (1.0), disodium hydrogen phosphate (0.3), potassium dihydrogen phosphate (0.2), sodium carbonate (0.1), calcium chloride((0.01), ferrous sulfate septahydrate (0.005) and manganous sulfate (0.002). Then 638 mg. of molecular sieve Linde 5A is added containing 50 mg. of n-hexadecane. A second 250 ml. bottle is similarly prepared.

To one is added a small amount, for inoculating purposes, of *Nocardia salmonicolor*, strain 107-332; no microbes are added to the second system which acted as the control. The bottles are then shaken, with the bottle loosely stoppered with a cotton plug so the air is available, for 120 hours at 30°C. and during the shaking period, samples are withdrawn aseptically from the inoculated system, and plating on nutrient agar plates is effected to follow cell growth.

The increase in the number of cells is rapid and significant. The cells appear as though they are growing on a water-soluble substrate.

In the control, the n-hexadecane is recovered unchanged. In a second control all the ingredients were present except the molecular sieve. In this instance the microbes and the hydrocarbon become clumped together; many cells were surrounded by hydrocarbon particles and were out of contact with the aqueous phase. Accurate sampling was impossible due to the clumping, but growth definitely was very much slower.

Emulsification by Nonionic Surface Active Agents

I. Takeda, T. Iguchi, K. Tsuzuki and T. Nakano; U.S. Patent 3,658,647; April 25, 1972; assigned to Asahi Kasei Kogyo K.K., Japan have found that yeast cells can be formed with stability and efficiency by continuously feeding a medium consisting of hydrocarbon, inorganic nitrogen, inorganic salts and organic source of nitrogen to which the fermentation waste liquor previously used and a small amount of a nonionic surface active agent are added into an emulsification tank.

Formation of the hydrocarbon emulsion is mainly effected in the first tank, i.e., the emulsification tank and the cells formation conducted in the second tank, i.e., the main fermentation tank with the emulsified fermentation liquor passed thereto from the first tank.

By the use of continuous fermentation carried out in such a manner, fermentation products other than the mycelia, such as amino acids, organic acids and vitamins, can be formed by the abovementioned species with a high efficiency.

The process is characterized by addition to the medium in the first tank of the fermentation waste liquor previously used, preferably at a ratio from 5 to 75% by volume based on the entire medium, and further, of a small amount of a nonionic surface active agent such as polyoxyethylene sorbitan fatty acid esters, sorbitan fatty acid esters or polyethylene glycol derivatives. The lighter liquor from centrifugal separation and sometimes the fermentation waste liquor prior to centrifugal separation or its hydrolyzate may be used as the fermentation waste liquor.

Continuous fermentation by the use of two-tank fermentation system enables feeding of hydrocarbon at a high concentration, thereby needing a smaller fermentation tank in which emulsion formation of the hydrocarbon requires a much shorter period of time as compared with the low concentration feeding. As noted, the process has a feature in the provision of an emulsified tank in addition to a main fermentation tank unlike prior processes. Efficiency of consuming hydrocarbon is very high. Generally, it is known that the yeast is formed from hydrocarbons, particularly from straight chain hydrocarbons at a ratio of 1 to 1 by weight. However, in this process the former is obtained at a ratio higher than 1:1 from the latter, and the fermentation products are obtained at a higher efficiency than those in the prior processes.

Example 1: Continuous fermentation of a yeast was carried out in two fermentation tanks containing media of the following compositions. The yeast used was *Candida petrophilum* ATCC 20226.

Medium in the First Tank

Straight chain paraffins containing from 14 to 17 carbon atoms (purity 98% or higher)	20 ml.
Ammonium sulfate	0.3 g.
Urea	0.3 g.
Monobasic potassium phosphate	0.1 g.
Dibasic potassium phosphate	0.1 g.
Sorbitan monolaurate	0.002 g.
Magnesium sulfate	0.01 g.
Ferrous sulfate	0.005 g.
Corn steep liquor	0.2 g.
Tap water	100 ml.
pH 5.5	

Medium in the Second Tank

Ammonium sulfate	0.3 g.
Monobasic potassium phosphate	0.1 g.
Dibasic potassium phosphate	0.1 g.
Magnesium sulfate	0.01 g.
Ferrous sulfate	0.005 g.
Vitamin B_1	100 γ
Tap water	100 ml.
pH 4.5	

The medium in the first tank contains 20% by volume of the hydrocarbon and the medium in the second tank contains no hydrocarbon. The fermentation liquor was transferred from the first tank to the second tank for effecting the continuous cultivation with the residence times under the conditions set forth below.

In the first and second tanks respectively in capacities of 8 and 40 l. were placed 4 and 20 l. of the medium respectively. The mycelia in the second tank after completion of the fermentation were removed by centrifugal separation and the lighter liquor was passed to the first tank to prepare a mixture of the lighter liquor and the fresh medium at a ratio of 4:6 by volume for the fermentation therein. The residence time in the first tank was 3 hours and the fermentation in the second tank was completed in 4 hours.

The fermentation conditions in the first tank were 30°C., pH 5.0 to 6.0 adjusted with aqueous ammonia, 400 rpm, and 100% of aeration. The second tank was controlled under such conditions as 33°C., pH 4.5 to 5.0, 400 rpm and 100% of aeration. The fermentation liquor thus produced was subjected to centrifugal separation by means of a Westfalia separator into lighter and heavier liquors. To the heavier liquor was added a sorbitan aliphatic acid ester at a ratio of 0.005 g. per 100 ml. of the fermentation liquor and the mixture was heated to 50°C. followed by centrifugal separation.

The mycelia thus separated were washed with water and dried over a drum dryer. The yield was 108 g. of the yeast mycelia on a dried basis per 100 g. of the hydrocarbon added. Efficiency of the isolation of the mycelia by the use of a Westfalia separator was 92%, which enabled use as they were in the animal feed as a protein source.

Example 2: Torulopsis petrophilum ATCC 20225 was cultivated by the process entirely identical with that in Example 1. A surface active agent polyethylene glycol lauryl alcohol ether was continuously added to the first tank at a concentration of 0.002%. Continuous fermentation was conducted using as the second tank a Waldhof type tank of the same capacity as in Example 1. Employing the conditions of fermentation and treatment after separation of the cells entirely identical with those in Example 1 resulted in the dried cells in a yield of 106 g. per 100 g. of the hydrocarbon added.

Water-in-Oil Emulsions

R.L. Gorring; U.S. Patent 3,642,577; February 15, 1972; assigned to Mobil Oil Corporation has demonstrated a method for growing microorganisms by incubating them in a culture mixture in the form of a water-in-oil emulsion where the oil is a hydrocarbon. The mixture is made up of aqueous mineral salt nutrient solution, a hydrocarbon as the sole source of carbon for energy and growth, and oxygen, and enough hydrocarbon is present to form a water-in-oil emulsion.

The hydrocarbon is one that is in the liquid phase at incubation temperature so as to be able to form a water-in-oil emulsion.

Aliphatic hydrocarbons are preferred, and these may be saturated or unsaturated, straight or branched chain hydrocarbons having up to 40 or more carbon atoms. Saturated straight chain hydrocarbons having up to 20 carbons are particularly desirable.

Cyclic hydrocarbons, comprising aromatic and alicyclic compounds, are also of use, including alkyl-substituted cyclic compounds having one, two, or more alkyl substituents each of any suitable length, chain configuration, and degree of saturation, and in which the cyclic moiety is aromatic or cycloparaffinic.

The hydrocarbon may be in the liquid phase not only by having a suitable melting point but also by being dissolved in a suitable solvent. The hydrocarbons are those which are normally liquid at incubation temperature. However, other useful hydrocarbons are those which are normally gaseous at incubation temperature, such as methane, ethane, propane, butane, and other C_3 to C_5 hydrocarbons.

These gaseous hydrocarbons may be dissolved in a normally liquid hydrocarbon, such as a petroleum fraction in the gasoline or kerosene boiling range, such as alkane like octane, nonane, decane, etc.; or they may be dissolved in any other conventional solvent which is inert in the process and nontoxic to the cells. Normally solid hydrocarbons may be used as the source of carbon by dissolving them in a hydrocarbon solvent.

The proportions of the aqueous mineral nutrient (which includes water and dissolved salts) are such as to provide the water-in-oil emulsion, the hydrocarbon forming the continuous phase. The emulsion is preferably formed and maintained by agitation of the culture mixture, and suitably may comprise at least 66%, and going up to 80 or 90% or more, of hydrocarbon, the balance being aqueous mineral nutrient, volume basis.

According to this preferred method of forming the emulsion, one or more emulsifying agents are present, either being produced in the culture mixture during the fermentation, or being added. Emulsifying agents added per se may be of ionic or nonionic character suitable to promote a water-in-oil emulsion.

A water-in-oil emulsion may be formed and maintained at lower hydrocarbon proportions, extending from about 66% down to 35 or 40%, balance aqueous mineral nutrient, volume basis, and at these proportions it is preferred to add one or more emulsifying agents.

It is feasible to add to the culture mixture a substance which takes up and releases oxygen, preferably one that is superior in this respect than the hydrocarbon. Such substance or compound can then take up oxygen from the introduced air and at some subsequent point in the flow of the culture mixture release at least a part of the oxygen to the cells. In this way a greater amount of oxygen may be presented to the cells for utilization by them.

A compound of interest for this purpose is hemoglobin, a water-soluble material comprising a complex of an iron-containing pigment portion, called heme or hemin, with a protein called globin. Hemoglobin has the ability to combine loosely with molecular oxygen, forming oxyhemoglobin, but the union is loose and can be dissociated under relatively simple conditions, such as a change of partial pressure of oxygen or of pH, forming oxygen and reduced hemoglobin. Thus, the hemoglobin can pick up oxygen in a region of higher oxygen partial pressure and release it in a region of lower oxygen partial pressure.

The hemoglobin complex may be separated into heme and globin. The heme is capable of combining with various other proteins, like albumin, and with bases like pyridine, nicotine, ammonia, etc., yielding hemopyridine, hemonicotine, etc. Collectively such compounds are called hemochromogens. They react with oxygen similarly to hemoglobin, and are of similar use.

Example: In a batch run, a culture mixture was prepared comprising 6.5 l. n-hexadecane, 3.0 l. of fresh aqueous mineral nutrient and 250 mg. *Brevi-bacterium sp.* The mixture had a pH of 7.0. It was charged to an airlift fermentor, of the type illustrated, of about 10 l. capacity, heated to 36°C., and air bubbled through at a rate of about 1.8 cu. ft./min. As growth took place, the pH of the mixture fell to 6.0, at which point it was maintained by addition of ammonium hydroxide. Distilled water and n-hexadecane were added periodically to replace losses by evaporation.

The concentration of the cells was measured periodically by means of optical density determinations made on total or complete samples of culture mixture. One cc of fermentor sample was diluted in 8 to 10 cc of 2-propanol, centrifuged, then redipersed in water in preparation for a density measurement. Centrifugation of each sample showed the liquid phase of the same to be composed of ⅔ oil and ⅓ water or aqueous medium. Staining with a water-soluble dye (Nigrosin) revealed that the liquid phase was a water-in-oil emulsion.

At all cell concentrations it was found that the aqueous phase of the withdrawn samples was able to separate from the hydrocarbon phase in about 20 seconds, a phenomenon not observable in the case of a culture mixture comprising an oil-in-water emulsion. Owing to this rapid separation of phases, a foam-free aqueous phase was formed which contained most of the cells; and over it was a hydrocarbon phase. After about 10 minutes, a distinct settling-separation of the cells in the aqueous phase was noticed, with no dispersed oil droplets present.

Optical density was measured by testing samples of culture mixture for the adsorption of visible light rays of a wavelength of 400 millimicrons (0.4 micron) in a colorimeter. The following table shows the increase in cell growth at periodic intervals throughout the run.

Time of Run, hr.	Cell Concn. g./l.	Time of Run, hr.	Cell Concn. g./l.
1	0.15	11	5.9
2	0.25	12	7.1
3	0.35	13	7.8
4	0.5	14	7.95
5	0.65	15	8.05
6	0.9	16	8.1
7	1.3	17	8.15
8	2.2	18	8.25
9	3.3	19	8.3
10	4.3	20	8.3
		25	8.4

Cell growth increased continuously throughout the run. The increased growth was sharp during the period from the seventh to the thirteenth hour, the concentration increasing from 1.3 to 7.8 g./l. Plotting the data for a concentration-time curve shows this sharp increase as an almost vertical portion of the curve. Growth remained continuous but leveled off during the period from the thirteenth to the twenty-fifth hours. A high of 8.4 g./l. was reached.

Atomizing a Liquid Phase Feedstock

B.M. Laine and C.L. Vernet; U.S. Patent 3,268,413; August 23, 1966; and U.S. Patent 3,510,403; May 5, 1970; both assigned to The British Petroleum Company Limited, England have described an improved process for cultivating a microorganism in the presence of a hydrocarbon feed stock in a fermenter which comprises forming a dispersion of the hydrocarbon feedstock in a liquid medium prior to feeding the feedstock to the fermenter, mixing the dispersion with the microorganism and thereafter feeding the mixture to the fermenter.

The small particles may be formed by atomizing a liquid phase feedstock into an aqueous phase or into a gas phase. The small liquid particles formed in gas phase must be transferred into a liquid medium. The dispersion of small particles in the liquid carrier medium may then be passed into the fermenter; alternatively the small liquid particles may be formed within the fermenter by direct atomization into the liquid material in the fermenter. The aqueous medium used for forming the dispersion outside the fermenter may consist of aqueous nutrient medium.

The fluid phase which is passed through the atomizer may be a mixture of hydrocarbon feedstock with either air or aqueous medium. A fluid phase is passed through the atomizer to form a suspension of fine liquid particles in a gas, the suspended material being caused to fall onto a moving liquid surface, for example of aqueous nutrient medium.

The small liquid particles may be formed by dispersing a liquid phase feedstock into an aqueous phase or into a gas phase under the action of ultrasonic waves. Preferably there is passed through an ultrasonic wave generating device a fluid phase which is a mixture of a hydrocarbon feedstock with either air or aqueous medium.

A fluid phase is passed through the ultrasonic wave generating device to form a suspension of fine liquid particles in a gas, the suspended material being caused to fall on to a moving liquid surface, for example of aqueous nutrient medium.

Example 1: Gas oil and the culture medium were pumped from two containers to the two branches of an injector by means of two piston pumps. Buffer reservoirs were interposed between the pumps and the injector to steady the back-pressure, which was regulated to lie in the range 14 to 16 kg./cm.^{2}.

The injector consisted of a stainless steel capillary tube (interior diameter 0.5 mm.) having the free end narrowed to leave an outlet hole of 0.05 to 0.08 mm. The other end of the capillary was welded to the two inlet tubes for gas oil and medium to form a Y.

The pumps were capable of control to give a flow which can vary between 0.3 and 1 l./hr. The mixture of gas oil and culture medium from the injector passed into a 5 l. continuous fermenter provided with an agitator which had a speed of 1,000 rpm, a ratio of the diameter of the blade:vat of 1:3 and had 8 blades.

The continuous fermenter also was provided with a ring of capillary air injectors, electrodes for pH control (pH maintained at 4), double external cladding to keep the temperature constant at 30°C. by water circulation, and a system for drawing off liquid by means of a piston pump.

For injection of the mixed gas oil and medium as a fine dispersion using the system described above, the capillary point of the injector may be mounted either in the body of the liquid (preferably beneath the agitator) of above the surface and pointing towards the bottom of the fermenter. The materials employed were *Candida lipolytica* as the yeast stock, Iraq heavy gas oil (percent by weight n-paraffins of 12) as gas oil source and a pour point of +17°C. Aqueous mineral nutrient medium was made up as follows.

	Grams
Diammonium phosphate	2
Potassium chloride	1.15
Magnesium sulfate·$7H_2O$	0.65
Zinc sulfate	0.17
Manganese sulfate·$1H_2O$	0.045
Ferrous sulfate·$7H_2O$	0.068
Yeast extract	0.025
Tap water	200
Distilled water q.s. 1,000 ml.	

The air supply was filtered and compressed.

To demonstrate the significance of using the pressure injection system, the following experiments were carried out:

(1) Culture with feed of gas oil and medium in separate circuits without previous atomization.

(2) Culture with atomization of the mixture of gas oil and medium by means of the system of capillary compression injection with capillary submerged.

(3) Same as (2), capillary above liquid (air injection).

The following results were obtained.

	Experiment		
	(1) No Atomization	(2) Atomization in Liquid	(3) Atomization in Air
Volume of liquid in fermenter (l.)	3	3	3
Dilution rate	0.07	0.10	0.11
Air flow, v./v./hr.	70	70	70
Liquid flow, l./hr.	0.21	0.3	0.33
Gas oil, g./l. of feed	180	180	180
Cellular density, g./l.	5	5	5.5
Average size of gas oil globules, microns	100	10	6
Cell division time, hr.	10	7	6.5
n-Paraffins consumed, percent	41	40	43

It will be noted that dispersion by capillary injection appreciably lowers the cell division time while preserving the same degree of dewaxing, that is, the rate of production of yeast and of dewaxed gas oil is increased without altering the degree of dewaxing.

Example 2: This example illustrates the use of feedstock formed by the dispersion of solid paraffin wax. The solid substrate (hard gatsch or solid paraffin wax) was melted in a container and injected as a fine dispersion in air into the culture medium in the fermenter, using an air atomizer consisting of a capillary tube mounted in a concentric air inlet tube. The injection can be made either into the body of the liquid or above the surface of the liquid. The fermenter employed was a 5 l. fermenter as described in Example 1.

The materials used were *Candida lipolytica* as the yeast stock; Dunkirk hard gatsch (MP 60°C.); paraffin wax (MP 60° to 70°C.) and a heavy Iraq gas oil which had a pour point of +17°C. and contained 12% by weight normal paraffins.

The aqueous mineral nutrient medium was as described in Example 1. The following experiments were undertaken with the object of making a finely divided solid substrate, rich in or enriched with normal paraffins, available to the yeast.

(1) Culture with ordinary heavy gas oil without preliminary atomization.

(2) Culture with dispersion of hard gatsch.

(3) Culture with heavy gas oil enriched with paraffin wax, atomized.

The results obtained are shown on the following page.

	Experiment		
	(1) Gas oil not Atomized	(2) Gatsch Atomized	(3) Gas Oil Plus Wax Atomized
Volume of liquid (liters)	3	3	3
Dilution rate	0.10	0.10	0.10
Air flow v./v./hr.	70	70	70
Liquid flow, l./hr.	0.3	0.3	0.3
Gas oil substrate, g./l. of medium	180	180	180
Cellular density	4	8.5	9
Size of substrate particles, microns	100	5	10
Dewaxing, percent by weight of paraffins in feed	32	69	73

The use of a fine dispersion of gatsch or gas oil enriched with solid paraffins produced by an air atomizer results in a marked increase in the cellular density for the same dilution. This shows that the availability of normal paraffins can be thus increased; the practical interest lies in the improved hourly production of yeast and dewaxed gas oil without altering the quality of the latter.

Example 3: This example illustrates the use of a feedstock which has been dispersed by ultrasonic vibration. Gas oil feedstock and the culture medium were pumped by a rotary pump to an ultrasonic vibration apparatus (fluid whistle) which is made up of a very fine steel plate mounted in front of a flattened tube which constitutes the flat jet by which the liquids enter.

The assembly was mounted in a cylindrical container forming a resonance chamber. The frequency employed was of the order of 20 to 22 kc./sec. The outlet of the whistle was connected to the fermenter by means of a flexible plastic pipe of 50 cm. length. The fermenter was a 5 l. continuous fermenter as described in Example 1 as were the materials used and process conditions.

The significance of the ultrasonic system is shown by the following experiments:

(1) Culture wihtout preliminary dispersion of the gas oil.

(2) Culture with ultrasonic dispersion of gas oil.

Data show that the ultrasonic dispersion system increases the cellular density for a given rate of dilution, i.e., increases the availability of the gas oil substrate to the yeasts, thereby enabling an increase in yeast production while increasing the dilution at the same cellular density which corresponds to a shortening of cell division time.

	Experiment	
	(1) No Dispersion	(2) Dispersion
Volume of liquid, liters	3	3
Aeration, v./v./hr.	70	70
Gas oil content of feed, g./l.	180	180
Dilution	0.1	0.1
Liquid flow, l./hr.	0.3	0.3
Cellular density, g./l.	4	9
Size of gas oil globules (microns)	100	5
Dewaxing, percent by weight of n-parafin in feed	32	74

OTHER TECHNIQUES

Dialysis to Improve Cell Growth

The process of *V.F. Coty; U.S. Patent 3,418,208; December 24, 1968; assigned to Mobil Oil Corporation* comprises incubating in a first zone a microorganism with an aqueous mineral salts nutrient in the presence of oxygen and, as sole source of carbon, a hydrocarbon. The cells grow in the culture mixture and produce various products, at least some of which are capable of inhibiting growth and which for this reason may be designated toxic metabolic products.

Continuously with the growth of the cells, the culture mixture, or at least the liquid portion containing the toxic products, is removed from the zone in a continuous stream, the amount of which is variable, and passed into a dialyzing zone for selective removal of products.

In this zone the mixture is moved along one side of a dialysis material while an aqueous mineral medium is disposed on the opposite side, and during such step products leave the mixture and pass through the dialysis material to the mineral medium, thereby purifying the mixture to an extent. The latter then leaves the dialysis zone and passes back to the first zone where cell growth is taking place. At the end of the process, a yield of cells is obtainable which is up to 8 or more times greater than that obtainable in the absence of the dialyzing step.

The method is applicable to any aerobic microbial species which is able to utilize a hydrocarbon as the sole source of carbon for energy and growth, including hydrocarbon-utilizing species of bacteria, fungi, yeasts, and molds, nonfastidious organisms are preferred, i.e., those which will grow in simplified salts media without necessity for additions of organic compounds.

The hydrocarbon may be chosen from a broad selection, including aliphatic, aromatic and alicyclic compounds of varying molecular weight and carbon configurations.

Crude oils, various petroleum fractions, residua, etc. are suitable. A preferred class of hydrocarbons comprises alkanes having up to 20 or 30 carbons, more preferably those that are gaseous or liquid at normal temperatures and pressures, and including straight and branched chain, saturated and unsaturated hydrocarbons.

In particular, normally gaseous hydrocarbons have an advantage in that any unused compound does not remain in the incubation mixture when the process is completed, but rather can simply be passed off to the atmosphere or recycled, thus avoiding separation problems. Gaseous hydrocarbons are also more economical; and they do not present problems of accessibility to the microorganism since, being gaseous, there is no question of the microorganism becoming imbedded in or occluded by the hydrocarbon. High concentrations of gaseous hydrocarbons are thus tolerated.

A further advantage is that in a continuous process, or in a batch system having a plurality of incubators, the gaseous hydrocarbon may be easily flowed through the continuous process, of from incubator to incubator in the batch system, and may be recycled.

By the term normally gaseous hydrocarbons there is meant those which are gaseous at temperatures up to about 55°C. Suitably, these compounds include hydrocarbons having 2 to 5 carbon atoms, more desirably 3 to 4 carbons, and particularly straight chain alkanes. In some cases methane is suitable.

Another preferred class comprises alkyl-substituted cyclic compounds having 1, 2, or more alkyl substituents each of any suitable length and comprising straight or branched chain, saturated or unsaturated radicals, and in which the cyclic moiety is aromatic or cycloparaffinic.

Example: A 2 l. fermentor containing mineral salt nutrients was innoculated with 15 cc of a culture of *Nocardia salmonicolor* 107-332. Then a mixture of about 90% air and 10% butane was bubbled through the culture mixture, during agitation of the latter, at a rate corresponding to 500 cc per minute of air, and the mixture was incubated at 30°C. The ensuing growth was followed by periodic optical density measurements, using a Bausch and Lomb colorimeter to absorb visible light rays of a wave length of 400 millimicrons. After one week, the culture stopped growing at a density reading of 11, corresponding to a yield of cells of 4 to 5 g./l., dry weight. The culture did not exhibit further growth despite additions of such materials as urea and phosphate salts.

Then a dialysis tubing was set up by immersing the same in a fresh aqueous mineral salts medium contained in a 40 l. tank. The tubing was made of regenerated cellulose and had an inflated diameter of ⅝ inch. Its length was about 10 feet and its volume about 1 liter. Slow agitation of the fresh medium was started, and then the culture mixture from the small fermentor was pumped through the tubing at a rate of about 90 cc/min., after which it was returned to the small fermentor. For reasons of safety, no pumping was done during times when the equipment was unattended. From optical density measurements, growth was immediately observed.

Thus, where the maximum cell weight before dialysis was observed to be 4 to 5 g./l., the cell weight one day after dialysis had started was about 6 g./l.; seven days after dialysis had started, the weight increased to 26 g./l.; and after sixteen days it increased to 46.85 g./l., corresponding to a density reading of 175.

During the course of the run, additional mineral salts medium was added from time to time to the large tank. The amount of such salt additions was substantially the same as the amount present in the fermentor at the very beginning of the experiment. When the density reading reached 41, corresponding to about 12.5 g./l. of cells, the air flow rate was increased to 1,000 cc/min. and the butane rate was increased to maintain a 10% concentration. This increased flow rate was maintained to the end of the experiment.

Reducing the Vapor Pressure of Volatile Hydrocarbons

R.I. Leavitt; U.S. Patent 3,474,001; October 21, 1969; assigned to Mobil Oil Corporation has described a method for increasing the growth of microorganisms on volatile hydrocarbons having up to about 8 carbon atoms by adding to the volatile hydrocarbon a monomeric hydrocarbon having more than 8 carbon atoms or a polymeric hydrocarbon. The use of the higher hydrocarbons effectively reduces the vapor pressure of the lower one, thus producing an increased growth response. The volatile aliphatic hydrocarbon which is the primary source of carbon for the microorganism is a saturated or unsaturated, straight or branched chain hydrocarbon having from 3 to 8 carbon atoms. These will include n-propane, n-butane, n-pentane, n-heptane, and n-octane as well as the branched chain members thereof. Also included are propene, butene, pentene, hexene, heptene, and octene.

The monomeric hydrocarbons will include hydrocarbons containing at least 9 carbon atoms, such as nonane, decane, dodecane, and tetradecane, as well as other straight and branched chain, saturated and unsaturated monomeric hydrocarbons containing up to 30 carbon atoms.

Polymeric hydrocarbons include such polymers as polybutene and polydecene. In general, any such monomeric or polymeric hydrocarbon which has a low volatility, is relatively inert (will not by itself adequately support microbial growth), and is miscible with the growth substrate, will be useful. The amount of hydrocarbon is that required to provide carbon for growth and energy and may be chosen for any desired growth period, although it may also be added from time to time to a culture mixture as may be necessary. It is desirable to add only the required amount so as to avoid separation problems at the end of the growth period.

Microbial syntheses by this method may be completed in times as short as one or two days. The incubation period may, of course, extend longer, but it is of interest to note that many microbial conversions, including syntheses, have in the past required periods of a week or two, or more, within which to produce appreciable growth.

The amount of monomeric or polymeric hydrocarbon will range from 0.1%, preferably from 2%, to amounts in excess of the aliphatic hydrocarbon present. While the quantity of monomeric or polymeric hydrocarbons is not critical, since even small amounts with the volatile aliphatic hydrocarbons will show some improvement, there are optimum levels which give the greatest increase in yields. The higher the number of carbon atoms in the aliphatic member the less of the higher member will be required.

On the other hand, the aliphatic hydrocarbons having low carbon contents, and corresponding high vapor pressures, will require an amount of monomeric or polymeric hydrocarbon which may exceed, sometimes substantially so, its weight.

Example 1: The hydrocarbon mixtures used in this example were prepared as follows. Six samples, each containing 3 ml. of n-octane, were prepared as shown in the table below.

	Moles Octane	Moles Tetra-decane	Vapor Pressure Octane
Sample:			
1	0.0185	0	18.48
2	0.0185	0.00038	18.10
3	0.0185	0.00116	17.39
4	0.0185	0.00193	16.73
5	0.0185	0.00270	16.12
6	0.0185	0.00347	15.56

A soil-isolated organism identified as a member of the genus Achromobacter was inoculated into seven 125 ml. Erlenmeyer flasks containing 50 ml. of the mineral salts nutrient referred to above. In one flask was placed 0.3 ml. of the above sample containing no tetradecane.

In a second flask was placed 0.3 ml. of the mixture containing 0.00038 mol of tetradecane, etc. In the seventh flask was placed a small amount (0.3 ml.) of tetradecane. The cultures, at a pH of 7.2, were shaken at 30°C. for 48 hours, and the growth was determined by optical density.

The resulting data are expressed below as optical density, and the relation between optical density and cell growth is approximately 1.0 g./l. of cells, dry weight, for every optical density of 4.0.

Sample	Optical Density
1	0.44
2	0.72
3	1.80
4	2.46
5	5.00
6	14.00
7	0

Example 2: Samples of n-octane were prepared as in Example 1, using the same incremental quantities of polybutene with a molecular weight of 625. Cells of the genus Achromobacter were grown as shown in Example 1. The effect on the growth and utilization of n-octane was essentially the same as observed for Example 1, as shown in the following table.

	Moles Octane	Moles Poly-butene	Vapor Pressure Octane	Optical Density
Sample:				
1	0.0185	0	18.48	0.21
2	0.0185	0.00038	18.10	0.28
3	0.0185	0.00116	17.39	0.36
4	0.0185	0.00193	16.73	1.04
5	0.0185	0.00270	16.12	3.70
6	0.0185	0.00347	15.56	6.60

No growth was observed on polybutene alone.

Carbohydrate Pretreatment Feedstock

The process of *B.M. Laine and J.A. Filosa; U.S. Patent 3,271,266; September 6, 1966; assigned to The British Petroleum Company Limited, England* concerns the cultivation of yeast by starting the growth of yeast on a carbohydrate starter feedstock followed by growth on a main feedstock containing a straight chain hydrocarbon.

It is a process which comprises, in a pretreatment stage, cultivating a microorganism in the presence of a pretreatment feedstock comprising a compound containing carbon, hydrogen and oxygen and thereafter, in a growth stage, cultivating the microorganism product so obtained in the presence of a growth feedstock comprising a straight chain hydrocarbon.

By operating in this manner the lag phase, which usually occurs when a microorganism is first cultivated on a hydrocarbon feedstock, is reduced in its effect or is eliminated.

It will be necessary, in both the pretreatment stage and in the growth stage to operate in the presence of a gas containing free oxygen and in the presence of an aqueous nutrient medium. This medium may be of different composition in the two stages. When operating in batchwise manner the two stages may be carried out in either the same or different fermenters. When operating in continuous manner separate fermenters or separate zones in the same fermenter will be necessary.

Usually the feedstock used in the pretreatment stage will be a carbohydrate, for example consisting of or contained in molassas, wort, malt extract, wood hydrolysis sugars or lyes obtained in the course of paper manufacture. A special pretreatment feedstock which may be employed in an extract of lipids obtained in the purification of yeasts by solvent extraction.

Preferably the hydrocarbon is C_{10} or higher. Suitably there may be used a hydrocarbon fraction derived from petroleum.

According to a preferred feature there is provided a process which comprises cultivating a microorganism in the manner as hereinbefore described in the presence of a petroleum fraction consisting in part of straight chain hydrocarbons and having a mean molecular weight corresponding to at least 10 carbon atoms per molecule, and in the presence of a gas containing free oxygen and separating from the mixture, on the one hand, the microorganism and, on the other hand, a petroleum fraction having a reduced proportion of straight chain hydrocarbons or which is free of the straight chain hydrocarbons.

The process is of particular value for the treatment of petroleum gas oil fractions which contain straight chain hydrocarbons in the form of waxes, since by the process, a gas oil of improved pour point is obtained while the waxes are converted to a valuable product.

Usually the straight chain hydrocarbons will be present in the feedstocks according to this process as paraffins; however, the straight chain hydrocarbons may be present as olefins; also there may be used a mixture containing straight chain paraffins and olefins.

It is an important feature that when cultivating yeasts in the presence of the feedstocks under conditions favoring the growth of the yeasts at the expense of the straight chain hydrocarbons, and other hydrocarbons, for example isoparaffins, naphthenes and aromatics are not metabolized or, at most, the proportion which is metabolized is very small.

Furthermore, unlike conventional chemical processes governed by the law of mass action, the rate of removal of straight chain hydrocarbons is not substantially reduced as the proportion of these hydrocarbons in the overall mixture of hydrocarbons decreases (except, of course, in the very final stages of removal).

Thus, when desired, the percentage conversion of straight chain hydrocarbons which is achieved can be maintained at a value approaching 100% without necessitating a very disproportionate expenditure of contact time to achieve small improvements. Furthermore, in the continuous process, this high percentage conversion can be achieved without resorting to the use of a long reaction path.

By the application of this process under conditions which limit the metabolization of the straight chain hydrocarbons it is possible to operate with the removal of only a desired proportion of these hydrocarbons.

Suitable feedstocks to the process include kerosene, gas oils, and lubricating oils; these feedstocks may be unrefined or may have undergone some refinery treatment, but will usually be required to contain a proportion of straight chain hydrocarbons in order to fulfill the purpose of this process. Suitably the petroleum fraction will contain 3 to 45% by weight of straight chain hydrocarbons.

Example: 40 l. of an aqueous mineral nutrient medium were introduced into a stainless steel fermenter having an effective capacity of 60 l. An inoculum was added. It consisted of 20 l. of a 24 hr. *Candida lipolytica* culture.

The inoculum was prepared using 30 g./l. of beet molasses, containing among other growth factors 0.04 mg./kg. of biotin, 50 mg./kg. of pantothenic acid and 5 mg./kg. of inositol, these three factors being particularly valuable for the growth of yeasts. When the inoculum was employed on the heavy gas oil there was no lag phase, the culture starting off immediately in the phase of maximum growth (cell division time 3 hours). The aqueous nutrient medium used had the following composition.

	Grams
Diammonium phosphate	2
Potassium chloride	1.15
Magnesium sulfate $\cdot 7H_2O$	0.65
Zinc sulfate	0.17
Manganese sulfate $\cdot H_2O$	0.045
Ferrous sulfate $\cdot 7H_2O$	0.068
Yeast extract	0.025
Tap water	200
Distilled water to 1,000 ml.	

Adaptation to Growth on Hydrocarbons

E.H. Wegner; U.S. Patent 3,510,401; May 5, 1970; and U.S. Patent 3,616,216; October 26, 1971; both assigned to Phillips Petroleum Company describes production of enzymatically active microorganisms for use in hydrocarbon conversion processes and recovery and use of the microorganisms. Enzymatically active microorganisms have been adapted to the conversion of hydrocarbon feedstocks by contacting the microorganisms with relatively volatile hydrocarbon during the growth cycle. In another aspect, the adaptation of microorganisms to hydrocarbon is obtained by contacting the microorganism during the cycle with a hydrocarbon having up to six carbon atoms. The microorganisms are preconditioned during the growth cycle with relatively volatile hydrocarbons.

A suitable microorganism to be adapted for hydrocarbon conversion is suspended in an aqueous nutrient medium wherein it is contacted with relatively volatile hydrocarbon under conditions which are sufficient to promote the consumption of at least part of the volatile hydrocarbon by metabolic action of the growing microorganism.

The essential characteristic of the hydrocarbons employed is that they can be readily volatilized at conditions of temperature and pressure that are not detrimental to the essential characteristics of the living microorganism. The hydrocarbons are those having from 1 to 6 carbon atoms per molecule. Such hydrocarbons can be rapidly and easily removed from the microbial cellular material following the completion of the desired growth.

After the desired degree of growth of the microorganism has taken place during which time the organism has become adapted to the nutrient hydrocarbon, the hydrocarbon gases are removed by any suitable means such as selective volatilization or flashing. Where relatively low boiling hydrocarbon is employed during the growth cycle or is found to be present subsequent to the completion of the desired period of growth, the system can be flashed at reduced pressure to effect the substantially complete separation of nutrient hydrocarbon and microbial growth product.

The procedure leaves the microbial cellular materials substantially unaltered in the aqueous nutrient solution and substantially free of oily deposits. The enzymatically active cellular material can then be separated from the nutrient liquid medium. Such separation can be effected by centrifugation and/or suction filtration with the consequent production of a clean, cellular material which is biologically active.

The material can be employed immediately to promote the oxidation of hydrocarbons under aerobic conditions or, if desired, the enzymatically acitve cellular material can be stored for considerable periods of time under suitable refrigeration. The microorganism thus recovered can be employed to inoculate liquid hydrocarbons under aerobic conditions to promote the conversion of such hydrocarbons by enzymatic metabolism in the absence of cell growth to yield the corresponding oxygenated derivatives of the feed hydrocarbon such as alcohols, aldehydes, ketones or carboxylic acids. The products obtained from such oxidation and the distribution of such products depends upon the nature of the feed hydrocarbon, the conditions of the oxidation and the extent to which the interaction is allowed to progress.

The following example serves to illustrate, in one particular case, the application of this growth process to the production of one suitable microorganism.

Example: A Nocardia culture was grown on a nutrient growth composition comprising the following materials dissolved in deionized water.

Substance	Weight Percent
$(NH_4)_2SO_4$	0.1
Na_2HPO_4	0.05
KH_2PO_4	0.05
$MgSO_4 \cdot 7H_2O$	0.02
Na_2CO_3	0.01
$CaCl_2$	0.01
$FeSO_4 \cdot 7H_2O$	0.0005
$MnSO_4$	0.0002
Yeast extract	0.1

The culture was grown in an atmosphere of 20 weight percent propane and 80% air for 7 days. (Cultures have been grown under these conditions for periods up to 21 days). Growth was slow but good. The supply of propane was discontinued and the cell mass was recovered by filtration from the liquid medium, frozen and stored. The cell yield was 1.2 weight percent of dried cells based on total oxidation medium after 10 days.

Any suitable species of microorganisms which are able to grow on a relatively volatile hydrocarbon material and which will produce the oxygenated derivatives of substrates in a nongrowth enzymatically active state are suitable for utilization in the process.

Although there are numerous hydrocarbons and mixtures that can be satisfactorily employed as the growth nutrient hydrocarbon, it is preferred to use relatively inexpensive mixtures of hydrocarbons having from about 10 to 30 carbon atoms per molecule.

It is also preferred that the mixture contain a substantial amount of aliphatic hydrocarbons or hydrocarbons having aliphatic side chains and desirably a substantial amount of paraffinic hydrocarbon structure. The concentration of inoculate microbial material employed in the cell growth cycle can vary considerably although it is generally desirable to inoculate the growth hydrocarbon with from about 0.01 to 1.0 weight percent active microbial cells based on hydrocarbon.

The temperature during the growth cycle can be maintained within the range of from 60° to 115°F. for a period of from about 1 to 48 hours. However, the conditions of temperature and reaction time during the growth cycle can vary considerably depending upon the rate and degree of cellular growth and/or hydrocarbon conversion that is desired.

Although the solvent hydrocarbons employed to remove residual nutrient hydrocarbon from the microbial growth product are very selective toward the growth hydrocarbons compared to the lipid materials within the cells, the selectivity is only relative and it is, therefore, unavoidable that, depending upon the severity of the washing operation, some minor amount of cell lipids will be removed along with the residual growth hydrocarbon.

The removal of these materials which are necessary to maintenance of active microbial cells can be substantially minimized and essentially eliminated through control of the degree and severity of the solvent extraction step. In this operation, it is preferred to employ from 0.1 to 10 parts by weight solvent hydrocarbon based on weight of cellular material at an extracation temperature of from 32° to 140°F. and a pressure of from 5 to 500 psia.

Fermentation Techniques

SPECIALIZED APPARATUS

Control of Residence Time in the Reactor

High protein food is obtained in a process described by *K.R. Guenther and M.B. Perkins; U.S. Patent 3,384,491; May 21, 1968; assigned to Esso Research and Engineering Company*. Bacteria are grown on hydrocarbons in a reactor wherein the bacteria are in the log phase and are maintained in a concentration at least as great as that for which the exponential increase in population increment is equal to the increment continuously withdrawn, the liquid residence time in the reactor being maintained at 1.5 to 4 times the minimum generation time for the reactor.

The process utilizes as feed a selected cut from a C_{11} to C_{30} paraffinic petroleum feed which has been purified to reduce the level of aromatics to below ½ weight percent, as the bacteria *Micrococcus cerificans* and optimum nonlimiting inorganic salts reaction mediums to obtain high selectivity to bacteria rather than by-product esters.

An excellent high protein feed supplement can be extremely economically prepared from cheap petroleum feeds by this process. The first and most important requirement of this process is the use of critically short liquid residence times in the reactor. By utilizing liquid residence times of 1.5 to 4 times, preferably 2 to 3 times the minimum generation time for the particular organism, e.g., *Micrococcus cerificans*, a large shift in selectivity to bacteria occurs as follows.

(1) 2.7 times minimum generation time:

77 lbs. n-paraffin + 89 lbs. O_2 + 22 lbs. inorganic salts, NH_3, etc.
→ 100 lbs. dried cells + 62 lbs. CO_2 + 26 lbs. H_2O

(2) 5.3 times minimum generation time:

89 lbs. n-paraffin + 127 lbs. O_2 + 22 lbs. inorganic salts, NH_3, etc.
→ 100 lbs. dried cells + 97 lbs. CO_2 + 41 lbs. H_2O

"Minimum generation time" is defined as the minimum time for the bacteria concentration to double occurring in the log phase (i.e., straight line on a log plot) of batch fermentation conducted under optimum conditions, i.e., temperature pH and nonlimiting (including use of excess agitation) supply of hydrocarbon, O_2 and optimum aqueous salts medium to the bacteria.

If a growth promoter is used in the continuous process the minimum generation time referred to above should include supply of the promoter. This minimum generation time for the particular bacteria may be determined from the literature or conveniently empirically determined in the laboratory preferably using a 1,500 rpm paddle stirred batch fermentor.

With respect to the surprising nature of the improvement obtained by limiting liquid residence time it would not be expected that short residence times would improve selectivity nor would one skilled in the art ordinarily attempt to use such short liquid residence times in the absence of the discovered improvement in selectivity. Thus:

(1) it is theoretically impossible to use in a continuous process liquid residence times less than 1.45 times the minimum generation time, i.e., in any reaction where the rate of formation of product is proportional to the amount of product present the minimum residence time for a continuous process is $1/1n2$ or 1.45 times the batch minimum generation time; and

(2) since the continuous process is subject to upsets and considerable variations from optimum conditions a margin of safety of at least 1, preferably 2, times the minimum residence time for a continuous process (viz 1.45 times minimum generation time)

would be applied to prevent the population of bacteria from being depleted, i.e., washed from the reactor. It is noted that the only cost for this safety factor is increase in size in the reactor and that this is a minor cost in the total cost of the process.

It is noted that it would not be expected that short residence times would appreciably improve selectivity and that the commercial advantages obtained by utilizing short liquid residence times are extremely large as follows.

(1) *Cooling Costs* – All bacteria growth processes are highly exothermic, e.g., 5,000 to 10,000 Btu/lb. cells and are conducted at temperatures near room temperature. Therefore, refrigeration rather than water cooling is ordinarily required. The cost of this refrigeration in a typical process where short residence time is utilized [equation (1)] is approximately 40% of the costs associated with fermentation (i.e., total costs composed of raw material costs plus fermentation costs plus product recovery costs). With longer residence times [equation (2)] this cost is increased approximately 60% (i.e., proportional to the additional amount of oxygen utilized to form CO_2 and water.

(2) *Power Costs* – All the present bacteria growth processes involve hydrocarbons dissolved in low concentration in an aqueous mineral salts growth medium. Since oxygen has very low solubility in water, very large power requirements (agitation and blowers) are necessary to obtain transfer of oxygen to the microorganism. These power requirements are not reduced by expedients such as emulsification although they can be reduced by using oxygen-enriched air rather than straight air. However, the cost of providing oxygen-enriched air is quite substantial. In any event utilizing short residence times [equation (1)], power costs are approximately 40% of fermentation costs and these costs would be increased by about 40% or more by utilizing longer residence times [equation (2)].

(3) *Hydrocarbon Costs* – Hydrocarbon costs are, of course, reduced 15% by utilizing short reaction times. However,

these costs are small compared to the costs referred to above.

(4) *Reactor Equipment Costs* – The size of the reactor is also reduced 15% by utilizing short reaction times. However, this cost improvement also is small compared to (1) and (2) above.

Eliminating essentially all aromatics from the feed permits an extremely economic recovery of the bacteria since residual unconverted hydrocarbon feed and product esters, etc., retained on the bacteria, have been found not to poison or to be otherwise detrimental to animal growth. These products retained on the bacteria are, however, of no advantage and are intentionally minimized by the preferred techniques.

The process will be more clearly understood from a consideration of the accompanying drawing, Figure 4.1, describing a preferred apparatus utilizing a preferred C_6 to C_{30} feed containing aromatics. Referring to the drawing, hydrocarbon feed is fed through line **1** to feed purification **2** where non-normal materials and impurities are removed through line **3**. This feed purification is preferably an absorption of the normal hydrocarbons, preferably paraffins, by 5A molecular sieves followed by desorption and then a cleanup of the desorbed normal hydrocarbons with 13X sieves or silica gel to absorb remaining impurities including aromatics.

The purified normal paraffins are passed through line **4** to normal paraffin storage **5**. The normal paraffins are supplied through line **6** to reactor **7** containing the bacteria and stirred by a stirrer **8**. Phosphoric acid is supplied through line **9**, inorganic salts are supplied through line **10** and water is supplied through line **11** to mixing tank **12** equipped with a stirrer **13**. The mixed salt solution is supplied through line **14** to the reactor. Ammonium hydroxide is also supplied to the reactor through line **15**.

Air or enriched air is supplied to the reactor through line **16**. Air CO_2 and water vapors are vented to the atmosphere from the reactor through line **17**. A mixture of product cells, salt solution and small amounts of unreacted hydrocarbon is continuously removed through line **18** and a part of this mixture is recycled through line **19**, cooling tower **20** and line **21** to provide cooling in the reactor.

Additionally or alternatively, a cooling coil **22** in the reactor may

FIGURE 4.1: PRODUCING HIGH PROTEIN FEED SUPPLEMENTS FROM HYDROCARBONS

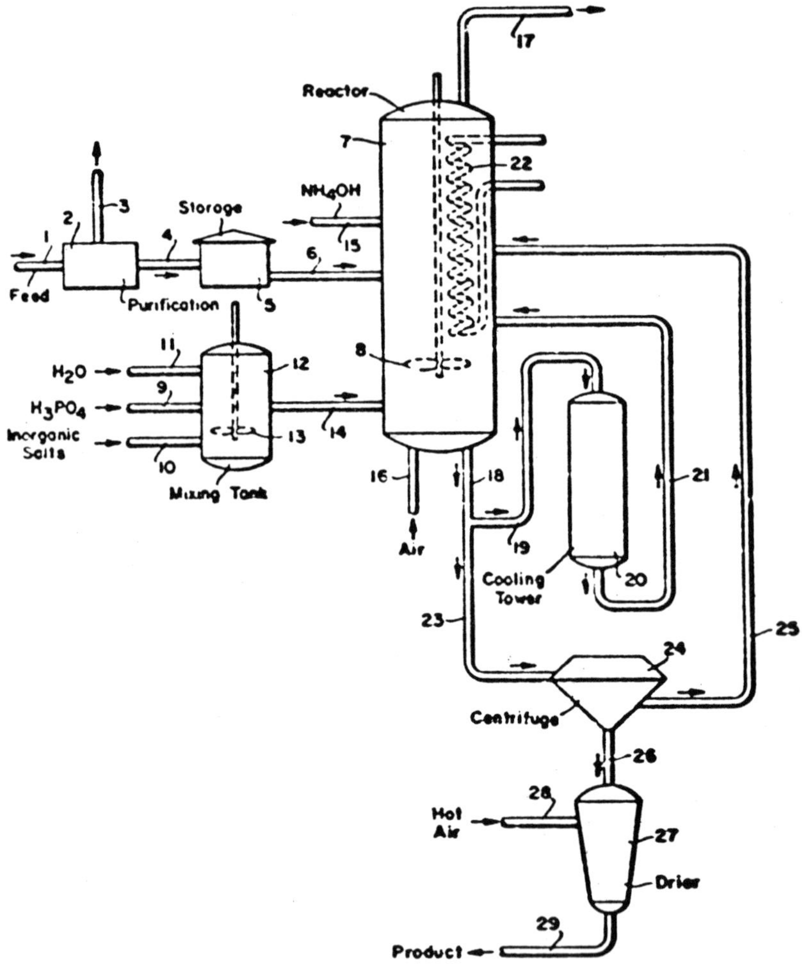

Source: K.R. Guenther and M.B. Perkins; U.S. Patent 3,384,491; May 21, 1968

also be employed. The remainder of the product mixture is supplied through line **23** to centrifuges or settling **24**, where product cells are separated. Supernatant liquid is withdrawn through line **25** and recycled to the reactor. Separated concentrated cells are passed through line **26** to dryers **27**, preferably spray dryers, utilizing hot air supplied through line **28**. Dried product cells are removed through line **29**.

Preferably the centrifuge separation is such that the weight percent dry cells (not including intracellular water) in the paste from the centrifuge is 7 to 28 weight percent dry cells, more preferably 10 to 15 weight percent dry cells. Such a separation can be economically effected and due to the critical reaction conditions the final product cells after drying will contain only 7 to 10 weight percent total fat of which about 75% is esters and 5 to 9 weight percent inorganic salts. Thus, the losses of the latter which are not particularly advantageous in the animal feed are not excessive.

It is noted that instead of the centrifuge separation other methods such as filtration or settling with or without filter aids or coagulants may be used to make the initial separation of cells and, of course, various drying procedures may be used. In any event it is noted that this procedure is extremely economic in that no special complex separations and surfactant washing to remove oil and by-products are required to obtain a final product having extremely high protein content and no impurities detrimental to animals or humans.

Example 1: Criticality of Residence Time — Micrococcus cerificans was grown continuously in the following medium (in grams per liter).

Slack wax	10 - 20
Span 60 - Tweens 60	0.6 - 1.2
H_3PO_4	5
KCl	1
$CaCl_2$	0.5
$MgSO_4 \cdot 7H_2O$	0.2
$MnSO_4 \cdot 4H_2O$	0.2
$FeSO_4 \cdot 7H_2O$	0.2
NaCl	0.2

Concentrated NH_4OH added as needed to control pH at 7.

Above nutrients dissolved or dispersed in tap water to obtain desired concentrations.

The temperature of the reaction was maintained at 30°C. and the fermentation carried out in 7.5 liter New Brunswick fermentor with an operating capacity of 4 liters. The air flow rate and agitator speed were varied so as to obtain aeration efficiencies of from 3.5 to 7.0 millimols O_2 per liter per minute.

The residence time in the fermentor was varied from 1.3 to 4.0 hours and the effect of the variation is shown below.

Residence Time, hours	Selectivity, wt. percent dried cells based on slack wax converted to cells and CO_2
4.0	107
4.0	102
2.0	120
2.0	128
2.0	134
2.0	133
2.0	121
1.3	138

It should be noted that this improvement is great not only because of the increased selectivity but also because this 30% improvement also reduces both the O_2 requirement and the heat removal requirement by 50%.

This experiment was repeated using a pure normal paraffin feed, i.e., n-hexadecane and analyses for unreacted hydrocarbon were accomplished by means of vapor fractometry. In this case it was possible to calculate selectivities as net percent product formed per unit hydrocarbon consumed. These results are shown below.

Residence Time, hours	Selectivity, wt. percent dried cells based on hydrocarbon consumed
4	105 ± 2
2	116 ± 6

Each of these values represents an average of from 5 to 6 experimental runs.

Example 2: Using the conditions described in Example 1, residence time 2 hours, *Micrococcus cerificans* was grown on n-hexadecane at levels of approximately 1 to 2 weight percent and air and pure O_2 as the sources of oxygen. When the hydrocarbon level was 1% there was very little effect of using pure oxygen. At a 2% hydrocarbon level, however, pure oxygen significantly improved hydrocarbon utilization as the following data show.

n-Hexadecane Conc., g./l.	O_2 Source	Selectivity, g. cells/g. C_{16} conv.	Hexadecane Conversion
9.3	Air	1.14	96
10.9	O_2	1.20	91
19.1	O_2	1.13	85
19.9	Air	1.12	73

A Two Step Process for Growth of Yeast and Bacteria

Yeast and bacteria being acclimated to the conversion of hydrocarbons are known to be effective agents for the removal of paraffinic hydrocarbons, particularly n-paraffins from mixtures with other hydrocarbons. It is also known that certain cultures produced by the enzymatic conversion of hydrocarbons have considerable nutrient value. Therefore, in applications requiring the removal of n-paraffin hydrocarbon from hydrocarbon mixtures, it is desirable to recover at least part of the protein-rich culture produced by the enzymatic conversion of paraffins.

In the process of *E.H. Wegner; U.S. Patent 3,489,648; January 13, 1970; assigned to Phillips Petroleum Company,* hydrocarbons containing, at least in part, n-paraffins, are contacted with yeast in a first fermentation zone to consume a substantial portion of n-paraffin. The unconverted paraffin is contacted in a second fermentation zone with or without intermediate separation with bacteria to further consume the unconverted paraffin.

High conversions are obtained through the use of this two step procedure due to the fact that partially converted hydrocarbon, i.e., hydrocarbon feed having the n-paraffin content substantially reduced in the first stage, is further reacted in a second stage whereby it is not diluted with fresh hydrocarbon feed being passed to the first fermentation zone having a substantially higher concentration of n-paraffins.

The protein product derived by the consumption of n-paraffins by yeast is more desirable than that produced by the consumption of similar hydrocarbon feedstocks with bacteria. However, certain species and strains of bacteria are more effective for consuming n-paraffins, particularly where paraffin concentration is relatively low.

Essentially any hydrocarbon can be treated by the process. However, as it is usually preferred that the reaction phase consist of a hydrocarbon-water emulsion, it is preferred that the hydrocarbon be substantially fluid at reaction conditions in order to facilitate such dispersion.

The hydrocarbons treated have from about 6 to 40 carbon atoms and the n-paraffins contained in the hydrocarbon feed have from about 10 to 25 carbon atoms. Broader ranges of molecular weight can be employed, although better economy is realized within these preferred ranges. Essentially all types of hydrocarbon mixtures comprising n-paraffin and/or branched and unbranched aliphatic, cycloaliphatic and aromatic hydrocarbons can be employed.

The n-paraffin in the hydrocarbon feed, regardless of relative concentration, will be converted, at least in some degree. However, where the production of high quality protein feed supplement is desired, relatively concentrated n-paraffins having essentially no other hydrocarbons or only minor amounts are used.

The degree of conversion realized will, of course, depend upon the concentration of microorganism in each fermentation zone as well as the concentration of n-paraffin in each instance. It should be realized, however, that these factors are only matters of degree and that some conversion will result, regardless of the relative concentration of microorganism and n-paraffin.

However, for the most economical operation, it is preferred that the concentration of microorganisms be within the range of from about 1 to 3.5 weight percent in the first fermentation zone and from about 1.5 to 4.0 weight percent in the second fermentation zone. The relative proportions of hydrocarbon-water phase can also vary considerably, although it is preferred that the weight ratio of hydrocarbon to water be within the range of from about 2:100 to about 20:100.

In operation the effluent from the first fermentation zone can, if desired, be separated by methods well known in the art, such as filtration or centrifugation. Solvent extraction with numerous solvents such as low molecular weight alcohol, can be used for purification. Of course, where all the microorganisms are removed, additional inoculant material must be added continuously to the second fermentation zone along with the hydrocarbon medium effluent from the first fermentor.

The most convenient source for the culture feed to the second fermentation zone is, of course, the second stage growth product found in the effluent from the second fermentor. This enzymatically active material can be removed from the effluent of the second stage by the same methods as those described for the intermediate separation following the first fermentation zone. Where it is desirable to maintain enzymatically active microorganisms, that separation condition should not be so severe so as to damage cell structure or remove necessary cell lipides.

Where the quality of protein product produced in the first fermentation zone is of importance in the overall economics of the process and the second fermentation step is conducted in the presence of bacteria, the intermediate separation referred to above may be most desirable in order to avoid contamination of the preferred protein supplement with the less desirable products produced in the second stage. However, advantage in ultimate conversion is achieved by passing the entire effluent from the first stage to the second stage in such operations in that the concentration of active microorganisms in the second stage, i.e., yeast and bacteria, is therefore increased.

In another embodiment where yeast is employed as the active microorganism in each stage, no intermediate separation is necessary and the microorganisms present in the effluent are quite effectively employed as inoculant material for the second fermentation zone. However, it may be desirable to recover at least part of the microorganisms produced in each stage before passage of the hydrocarbon effluent to the next fermentation zone; in which event, the intermediate separation described above can be employed.

In either event, a part of the growth product removed from the effluents of the several fermentors can be recycled to the same fermentation zone as inoculant material. However, this is usually not necessary, for growth of the microorganisms is usually rapid enough to supply all the microorganisms needed. It can be seen, therefore, that once the conversion is in progress, it is not necessary to provide active microorganisms as inoculant material from external sources.

n-Paraffin utilization of up to 99.5% can be achieved by the process. as compared to ultimate utilization of about 90% obtained under the same conditions in single stage operation. n-Paraffin utilization of a first fermentation zone will usually be within the range of from about 70 to 90% as compared to ultimate utilization of from about 99 to 100% in the second stage fermentor.

Example 1: An aqueous mixture containing 3.89 weight percent C_{12} to C_{15} n-paraffins was continuously passed to a first stage fermentor having a holding time of 6 hours. The first stage was operated at a temperature of 30°C. and a pH of 5.0. Air was passed into the reaction phase at a rate of 1.5 volumes per volume per minute. Ammonia was added as necessary to control pH, about 0.5 part per 1,000 parts per hour on a weight basis. Agitation was supplied by an impeller turning at 1,000 rpm.

The first stage effluent containing 3.03 weight percent yeast and 0.64 weight percent n-paraffin (83.5% first stage utilization with a yield of 93.2% based on n-paraffins consumed). This effluent was fed directly to a second stage fermentor having a holding time of two hours, a temperature of 30°C. and a pH of 5.0.

The second stage effluent contained 3.59 weight percent yeast and 0.2 weight percent n-paraffin representing an ultimate utilization of 99.5% of the n-paraffin and an ultimate yield of 92.8% (based on grams yeast produced per grams n-paraffin utilized).

Example 2: The feed to the first fermentation zone of Example 1 was passed continuously to a single fermentor having a holding time of 8 hours and operated at conditions identical to those of the fermentation zones of Example 1. The effluent contained 3.2 weight percent yeast and 0.4 weight percent n-paraffin representing 90% n-paraffin utilization.

Example 3: The effluent from the first fermentation zone of Example 1 was centrifuged to remove essentially all of the yeast contained therein and the filtrate was passed to a second fermentation zone under conditions identical to those of the second stage of Example 1 wherein the filtrate was contacted with bacteria species. The effluent from the second fermentation zone contained 0.05 weight percent n-paraffins, 0.63 weight percent bacteria representing a second stage n-paraffin utilization of 99.2% and an overall yield of 100% based on n-paraffins consumed.

Example 4: The feed to the first stage fermentation zone of Example 1 contained 12 weight percent of a waxy diesel fuel rather than pure n-paraffins. The pour point of this fuel was 35°F. and it contained 20.9% n-paraffins. Operating conditions were essentially the same as in Example 1. The first stage effluent which was fed directly to the second stage contained 2.3 weight percent yeast, and the diesel fuel pour point was reduced to 10°F. and the n-paraffin content to 7.8 weight percent. The effluent from the second stage contained 2.7 weight percent yeast and the diesel oil pour point was reduced to below 60°F. and the n-paraffin content to 2.6 weight percent.

A Horizontal Cylindrical Chamber

Conventional fermentation vessels are not generally satisfactory for the production of protein through fermentation of hydrocarbons because the fermentation takes place in a mixed hydrocarbon-aqueous phase and it is necessary to provide for substantially continuous stirring of the biological nutrient mixture.

Fermentors in the form of vertical tanks with a stirrer, foam breaker and cooling jacket have been proposed for use in the production of biological protein products from hydrocarbons. Foam production is often severe with such apparatus since the stirrer, often in the form of a propeller blade, must rotate at a sufficiently high speed to maintain a substantially homogeneous composition.

Such equipment is not well adapted to continuous processing because the mixture is completely homogeneous throughout and no satisfactory method is available for continuous feed and continuous removal of spent nutrient and products. Therefore, *P.F. Mako; U.S. Patent 3,594,277; July 20, 1971; assigned to Atlantic Richfield Company* has developed an improved fermentation apparatus primarily for use in hydrocarbon fermentation processes.

The apparatus may be described as a horizontal elongate cylinder having end closure members at the respective ends. An elongate shaft is supported concentrically through the length of the cylinder by journal bearings in the end closure members. The shaft carries a plurality of discs thereon for rotation with the shaft. Each of the discs is provided with a multiplicity of apertures. Means are provided for rotating the shaft. Inlet conduits provide means for supplying air, hydrocarbon, nutrient and buffer solutions at one end of the cylinder and, similarly, conduit means are provided in the end closure at the other end of the cylinder for removing gaseous products and the spent liquid products.

In operation, the liquid level is maintained in the cylinder by draining the liquid at a given point through the removal conduit. This provides a gas phase above the liquid phase. The shaft and the discs thereon are rotated at a sufficient velocity to maintain the liquid phase continuously in locally substantially homogeneous composition. That is, at a given point along the length of the cylinder the liquid phase will be substantially homogeneous and portions of the liquid phase will continuously be carried into the gas phase for contact with the oxygen therein and for carrying oxygen into the liquid phase. The liquid phase and the gas phase will not, however, necessarily be homogeneous throughout the length of the cylinder since, as the result of biological action, the nutrients and the air will be used up as it passes from the entry end of the cylinder to the exit end of the cylinder.

Figure 4.2a is a side elevational view of the apparatus of this process shown in cross section. Figure 4.2b is an end cross-sectional view taken substantially along lines **2–2** of Figure 4.2a in the direction of the arrows. Figure 4.2c is an end view taken substantially along lines **3–3** of Figure 4.2a in the direction of the arrows showing the inlet conduit system.

The preferred embodiment of the process comprises, in combination, a cylindrical member made up of an inner cylindrical shell **10** and an outer cylindrical shell **12** joined together by an annular member **14** at one end and a similar annular member **16** at the other end. The two cylindrical shells **10** and **12** joined together in the manner shown in Figure 4.2a form a cooling or heating jacket through which water of a desired temperature may be circulated by means of conduits **18** and **20**. The jacket may be partitioned into several sections, if desired, where it is desirable to maintain different portions of the cylinder at different operating temperatures.

End closures **22** and **24** are secured, respectively, by a plurality of bolts **26** and a plurality of bolts **28** to the annular members **14** and **16** of the cylinder construction. The entire cylinder assembly is supported by blocks **30** and **32**.

A very important facet of the process is the provision of an elongate shaft **34** which extends, in the preferred embodiment, from a journal bearing **36** in the closure member **24** through the length of the cylinder, along the axis thereof preferably, and through a journal bearing **38** in the other end closure **22**. The shaft may be rotated in the journal bearings by any desired means, such as a hand crank or the like. In an exemplary embodiment, a bevelled gear **40** meshes with a second bevelled gear **42**, the second bevelled gear being driven by a shaft **44** on a motor **46**. Desirably, the motor is of the variable speed type so that the speed of rotation of the shaft may be controlled.

FIGURE 4.2: PROCESS FOR FERMENTING HYDROCARBONS

a.

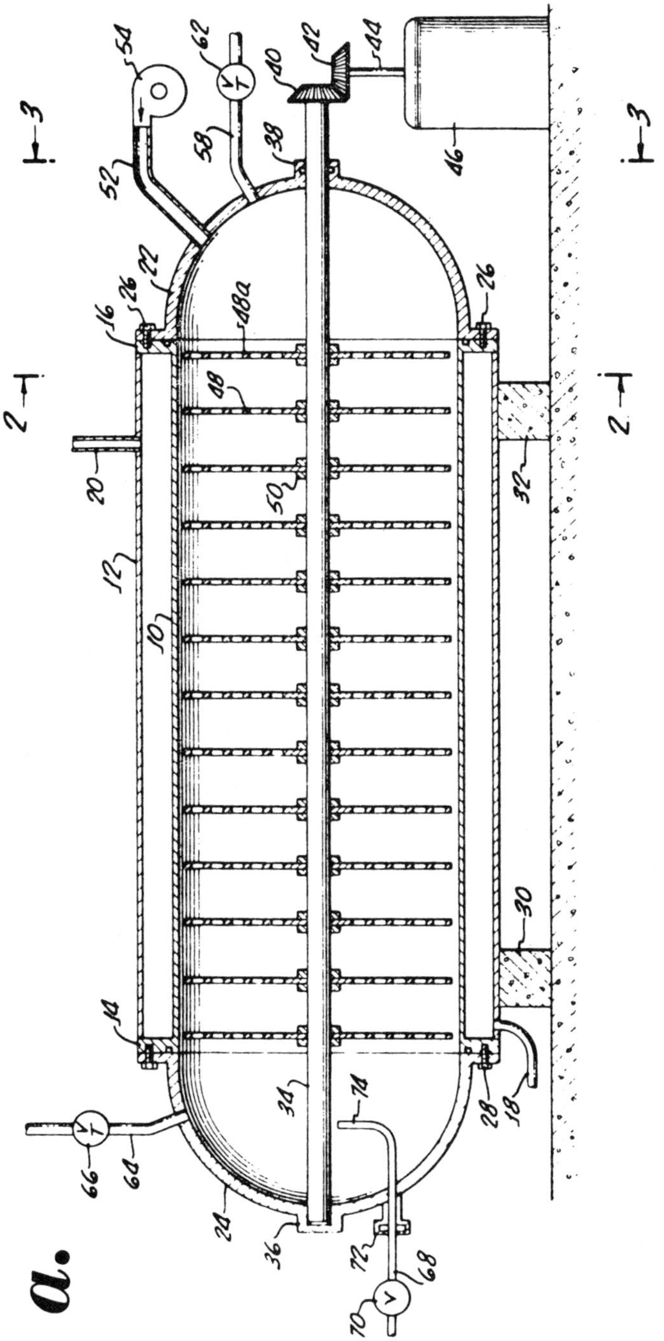

(continued)

FIGURE 4.2: (continued)

b.

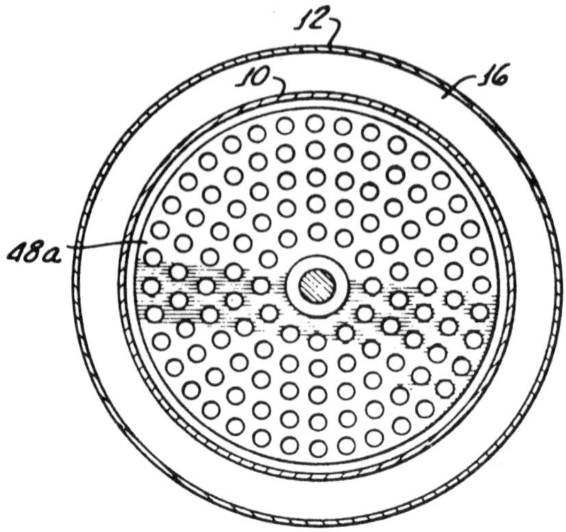

c.

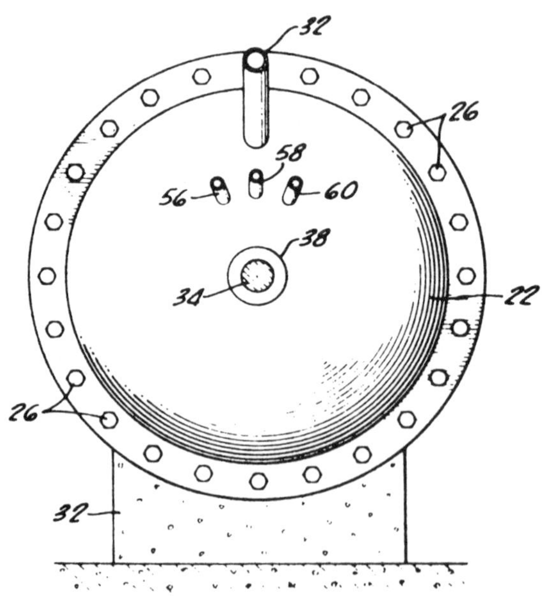

Source: P.F. Mako; U.S. Patent 3,594,277; July 20, 1971

A plurality of discs **48** are mounted by sleeves **50**, or equivalent mounting means, on the shaft **34** for rotation therewith. One of these discs, **48*a***, is shown in cross section in Figure 4.2a for purposes of illustration and is shown in side elevation in Figure 4.2b. Each of the discs is provided with a multiplicity of apertures extending therethrough. The apertures are preferably arranged circumferentially about the axis of the disc.

In the preferred embodiment, the disc is approximately the same diameter as the inner diameter of the cylinder such that the flow of the liquid in the cylinder is regulated to a degree by the construction of the disc and the speed of rotation thereof, as will be discussed in greater detail hereinafter. It is not necessary that the disc fit very tightly inside the cylinder but a very large space between the cylinder wall and the disc should be avoided since substantial portions of the liquid may flow longitudinally along the cylinder with little co-action with the gas phase if a free path of flow between the cylinder and the disc is provided.

The inlet end of the cylinder is provided with an air inlet conduit **52** which supplies air or other gas from a pump **54**.

Liquid inlet conduits **56**, **58** and **60** are also provided at the inlet end of the cylinder. Each of these conduits may be provided with a throttle valve **62**, as shown in conduit **58** in Figure 4.2a for metering desired quantities of liquids into the inlet end of the cylinder. Conventionally such liquids would include the hydrocarbon, biological nutrient solutions, and other buffer and nutrient solutions such as a dilute solution of ammonium hydroxide.

The exit end of the cylinder is provided with a gas outlet conduit **64** which may also be provided with a throttle valve **66** if desired. It will be apparent that the pressure inside the cylinder may be regulated by means of the throttle valve **66**. The effluent end of the cylinder is also provided with at least one liquid conduit **68** which may be provided with a cutoff valve **70**. One or more additional liquid conduits may be provided as desired.

The liquid effluent conduit **68** is preferably received through a packing seal **72**, or equivalent, for extending interiorly of the cylinder and is provided with a side arm **74** to form a generally L-shaped configuration. The conduit may selectively be rotated in the seal and moved reciprocally therethrough. This permits the interior end of the conduit to be positioned, by rotation, at the desired level for maintaining the liquid level in the cylinder.

Before proceeding to the process conditions proper, one additional important facet of the construction is worthy of note. Reference is made to the air inlet system conduit **52** and pump or blower **54**, and the air exit system, conduit **64** and the valve **66**. It will be apparent that the rate of air flow from one end of the cylinder to the other will be a function of the capacity of the blower **54**, the conduit sizes and the setting of the valve **66**.

While other means of controllably varying the air flow through the cylinder are available, such as varying the speed of the blower **54**, let it be assumed that the air flow through the cylinder is to be controlled by the setting of valve **66**. Thus, simply by adjusting valve **66** the air available to the biological system can be controlled.

An important ancillary benefit also results. It is well known that biological systems of the type under consideration are highly exothermic and that it is necessary to remove rather large quantities of heat. For this purpose, the water jacket is provided. Water jackets, in general, are not very efficient for heat removal in such systems because only a small portion of the body of the liquid is in contact with the wall of the cylinder at any given time. It would, of course, be more desirable to remove a substantial portion of the heat of biological reaction at the point of reaction.

This apparatus permits an approach to this ideal condition. As the air flows through the cylinder, there is an exchange of heat from the biological liquid medium with the air and there is a constant evaporation of the liquid into the flowing air stream. The liquid is made up from the feed system. Since the biological action requires oxygen, in the normal case, maximum biological reactivity occurs on the discs as they rotate through the gas phase and, consequently, the maximum heat production is in the gas phase. Since the gas (air) flow is controllable, the rate of heat removal by heat exchange and evaporative cooling can be controlled simply by adjusting valve **66**.

The process may be carried out as follows. First, the desired quantity of the selected hydrocarbon, nutrient solution, and nutrient buffer solution is introduced into the cylinder to provide the selected liquid level. The liquid level in the cylinder may be determined and maintained by the disposition of the exit conduit in the effluent end of the cylinder. Once the desired liquid level is attained, the temperature of the mixture is raised as desired by flowing water of the selected temperature through a water jacket. Air is introduced and flows along the upper portion of the cylinder and exits through a conduit.

During operation, the motor is energized and the shaft with the discs thereon rotates at the selected velocity. The rotational velocity must be determined according to the specific liquids being handled since the amount of the liquid which will be carried into the gas phase and the rotation necessary to provide a locally homogeneous phase will depend upon the viscosity of the aqueous and organic phases and upon the relative interfacial tension between the two phases.

As the discs rotate, however, they form a partial barrier to the flow of liquid along the length of the cylinder. Thus, the liquid phase is continuously carried into the gas phase, the gas phase is continuously intermixed into the liquid phase and the flow of the liquid phase is controlled by the rotation of the apertured discs so that circular layers of liquid are continuously carried into the air by the discs.

At the lower rotational velocities, the predominant flow will be linearly along the length of the cylinder. As the rotational velocity of the discs increases, however, the rotational flow of the liquid will tend to predominate. Thus, a plurality of locally homogeneous gas-aqueous-hydrocarbon zones may be formed.

As a result of biological action, however, the liquid flowing toward the effluent end of the cylinder will not be of the same composition as the entering liquid. Thus, the liquid flowing toward each successive disc will differ slightly from the liquid flowing toward the previous disc. A plurality of locally homogeneous biological reaction zones is thus provided in a single liquid container. This is an important advantage over homogeneous fermentation apparatus.

It may be desirable to provide a different ratio of gas-hydrocarbon-aqueous phase contact along the length of the cylinder. This is easily done simply by providing discs which have differing sizes of apertures and differing numbers of apertures therein.

For example, let it be assumed that it is desired to provide maximum air contact with the liquid phase and maximum intermixture of the liquid phase at the first stage and minimum air contact and liquid intermixture at the last stage of the reaction zone in the cylinder. In this case, the first disc would be provided with a very large number of comparatively small apertures therethrough whereas the last disc along the length of the cylinder would be provided with a small number of relatively larger apertures. A continuum of disc constructions along the length of the cylinder may obviously be provided.

Since the microbiological action is normally exothermic, once the reaction is initiated it may be necessary to substitute a cooling water of the desired temperature for the initial water used to raise the temperature to the desired reaction temperature. The reaction temperature, therefore, may be controlled by regulating air flow rate or the temperature of the water in a cooling jacket. A segmented cooling jacket may be provided by placing annular baffles at spaced points along the length of the cylinder in the jacket and providing additional inlet and outlet conduits.

A specific design may be as follows. The interior cylindrical diameter may be 10 inches and the length of the cylinder may be 60 inches. The outer cylinder, **12**, may have a diameter of 12 inches, thus providing a water jacket of approximately 1 inch thickness. Approximately 60 discs may be mounted along the length of a 1 inch shaft axially of the cylinder.

Each of the discs is between $^1/_{32}$ and ¼ inch in thickness and preferably between $^1/_{32}$ and $^1/_{16}$ inch in thickness and perforated in a plurality of circles about the axis through which the shaft extends with $^3/_{16}$ inch apertures. Apertures up to ½ inch in diameter may effectively be used, however. The discs are rotated as speeds between about 1 and about 300 rpm although higher velocities may be desired if larger apertures are used. Velocities between about 30 and 60 rpm are normally preferred.

The cylinder, including the space provided by domed end closures, contains approximately 300 liters. From about ¼ to ½ of the cylinder volume, and preferably about ⅓, is filled with liquid. The liquid comprises the hydrocarbon and aqueous nutrient solutions. Preferably about ⅔ of the cylinder is reserved for a gas phase. As the discs are rotated, the two liquid phases are intimately intermixed and are carried into the gas phase and, simultaneously, gas is carried into the liquid phase by means of the apertures in the discs.

The velocity of the liquid flow through the cylinder will be controlled by the rate of feed input, although local control of flow may be regulated by varying the speed of rotation of the discs.

Flow rates of the liquid through the cylinder may be controlled at any desired rate. Flow rates may range from about 0.001 to 10 l. pr. hr. and preferably about 0.001 to 1 l. per hr. The residence time of the liquid nutrient in the cylinder is, consequently, from about 10 hours to greater than 1,000 hours, depending upon the desired flow rates.

Often, the optimum residence time for maximum production of protein from hydrocarbon will control the residence time. Obviously, the size of the reaction cylinder and the associated equipment may be varied according to the desired residence time to produce protein at a commercially attractive rate.

In operation, a hydrocarbon such as C_{18} straight paraffin and an aqueous nutrient solution in the ratio of about 0.5 to 2.0 is introduced at one end of the cylinder and allowed to flow at a rate of about 0.1 l. per hr. through the cylinder. Air at a pressure from about 1 to 20 atmospheres, and preferably about 1.25 atmospheres, is maintained above the liquid phases. The liquid flows to the first disc where it is rotated into the gas phase for contacting the oxygen therein.

The liquid then flows successively to the second disc where it is rotated into the gas phase again, then to the third disc where it is similarly rotated into the gas phase, etc. along the length of the cylinder. The spent nutrient solution containing the protein produced is then removed from the cylinder. Fresh nutrient and hydrocarbon may continuously be introduced into the cylinder.

Continuous Microbial Process

The procedure of *R.A. Oppermann; U.S. Patent 3,591,455; July 6, 1971; assigned to Nalco Chemical Company* describes the utilization of a source of carbon and a separate source of nitrogen in a continuous system wherein by varying the choice of input carbon source and a selected initial batch of microbial agent, oxygenated or reduced hydrocarbon products are formed where the microbial agent is operated under nonproliferating conditions.

The process further contemplates the production of proteins and amino acids as the source of nitrogen where extra cellular nitrogen is provided under nitrogen-supplemental conditions, such as by an inorganic compound like ammonium sulfate. The system is continuous and carefully regulated so that even though carbohydrate product is desired from the source of carbon, a portion of the stream and the microbial particles are passed through the process to the source of nitrogen in a growth cycle before returning to the input source of carbon.

After both primary treatment stations in the process centrifuges separate the microbial agent from the medium, the desired commercial product may be further separated and recovered from the stream where optionally the preceding treatment station is live, i.e., where C or N cells are harvested and optionally commercial products are produced.

After each live separating station, the total microbial agent is split; a major portion is harvested and the remainder is passed along to the subsequent treating station of the opposite source (i.e., C or N). A portion of the effluent containing the medium is preferably recycled where the station is live and in a minority of cases a portion of the microbial agent may be recycled to provide inoculum.

Figure 4.3 is a block diagram teaching the major aspects, reading from left to right, carbon input, for example hydrocarbon, flows into **1** which is the carbon treating station containing a microbial agent selected to effectively introduce the desired results such as reduction or oxidation of hydrocarbon. As a startup, an initial or seed portion or bath of the microbial agent is prepared and placed in **1**. After this initial portion, in most cases due to features later explained, the microbial agent is sustained by continuous return for the nitrogen treating station **3**. Treating station **1** may be of a bed or tank containing agitation and oxygen input means.

The effluent from **1** passes to centrifuge **2** which may preferably be either of the dry bowl type or liquid-liquid centrifuge. A convenient bowl type apparatus is described in Perry's "Chemical Engineers Handbook", 4th ed., 1963, 19-88, Figs. 19 to 137. Alternatively separation may be by setting, precipitation, etc. Here the carbon rich microorganisms are separated (split, i.e., harvested and passed on) from the stream and if the desired extra cellular product is carbon rich, then this portion is further separated and recovered. Typical products are polysaccharides, polyhydroxybutyric acid, glycogen, etc.

In a continuous cycle, if carbon rich compounds such as above are desired, then approximately two-thirds of the cells are harvested.

FIGURE 4.3: FLOW CHART FOR CONTINUOUS MICROBIAL PROCESS

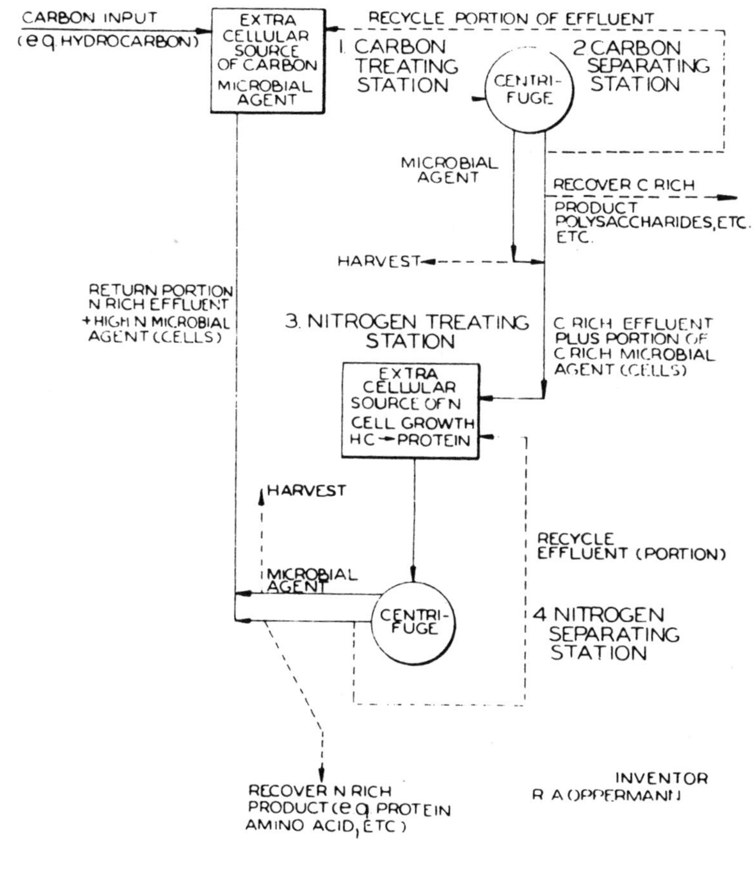

Source: R.A. Oppermann; U.S. Patent 3,591,455; July 6, 1971

In the event that the N station but not the C treating station is "live" or producing commercial compounds, then all the C rich cells are passed on the N treatment. These cells then go on stream to the nitrogen treating station **3** which provides an extra cellular source of nitrogen. Conventional agitation means may be present. Where nitrogen rich products such as proteins and amino acids are the main extra cellular products desired, station **3**, providing proliferation conditions, is where this reaction takes place.

The source of nitrogen may be organic or inorganic, but inorganic ammonium salts, such as the sulfate, chloride, etc. are preferred, for cost reasons and to avoid possible interference. Operable organic sources of nitrogen include proteins, enzyme-digested proteins, amino acids, yeast extract, asparagine and urea.

From station **3** a second centrifuge system **4** separates the nitrogen rich microbial agent from the effluent. If the nitrogen treating station **3** is live, i.e., designed in this process to produce commercial nitrogen rich products, then further separation and recovery is made of such nitrogen rich products as protein, amino acid, etc. by solvent extraction, ion exchange electrophoresis, etc. (not shown).

As was the case in the separation station **2**, the enriched microbial agent coming from separation station **4** is split. If N enriched products are desired in the process about two-thirds of the cells are harvested. On the other hand, where the primary product is C rich, all of the cells are passed on stream forward to the cell growth or C treating station.

The balance of the split enriched microbial agent from the separating station **4** not harvested may be recombined with the effluent from separation station **4** containing intermediates, aqueous or solvent remainders, and returned in a continuous manner to the subsequent carbon treating station **1** where C compounds are again accumulated.

Thus, a portion of all of the cells richest in N after separation at centrifuge **4**, are returned to the carbon treating station **1**, enhancing a continuous or cyclical source of microbial agent after provision for the initial startup or seed.

In general, after each primary treating station, i.e., for C and N, a separation station (e.g., centrifuge) separates the microbial agent (cells). Where the treatment station is live the microbial agent is split and a portion of the cells is forwarded downstream to the succeeding treating station. If carbon rich or nitrogen rich compounds are desired at this preceding treating station, then about two-thirds of the total microbial agent recovered is harvested. On the other hand, where the treating station is inactive for the required C or N product, then all of the microbial agent is forwarded downstream to the treatment station for the other (N or C) treating station.

In any case, the remainder of the cells or microbial agent recovered at the particular separation station is passed along to the next or subsequent treating station. For example, that portion of the cells from separation station **2** designed for nitrogen treating station **3** may conveniently be recombined with some of the effluent from centrifuge **2**. Where the treating station is live, i.e., where the desired product is produced at the preceding treating station, then the recombination of the microbial agent with the effluent occurs after further recovery of the desired product by separatory measures such as ion exchange preferential solvent or electrophoresis methods (not shown).

Where the treatment station is "dead" or not producing cells or extra cellular commercial product, then no product separation occurs and recombination of the split microbial agent with some of the effluent from the centrifuge may occur on the stream **2-3**.

Where the microbial agent or cells per se are the commercial product and the source of carbon and separate source of nitrogen are for substrate purposes only, then the desired carbon or nitrogen treatment station is considered live. Thus, after contact with the carbon source the microbes will be high in carbon compounds, as polysaccharides, polyhydroxybutyric acid, glycogen, etc., but low in nitrogen material. Therefore, where these carbon compounds are desired, about two-thirds of the cells are harvested at the carbon separating station as product and recovered. The balance is sent forward to the nitrogen treating station.

Where cells high in nitrogenous material are required, such as in the case where protein is desired as the product, then no harvesting occurs at the carbon separating station, but rather at the nitrogen separating station where about two-thirds of the nitrogen rich cells are harvested as products and the balance is sent forward to the carbon treating station.

Further, it is to be noted that a portion of the microbial agent from the centrifuge **2** designed for nitrogen treatment station **3** may be passed directly to this subsequent nitrogen treating station without recombination with the effluent as where it is desired that all or a portion of the effluent containing intermediates etc. be discarded. The same observation is true of a split of the microbial agent after separation in nitrogen separating station **4**, as in carbon separating station **2**. The split portion of the microbial agent, in an appropriate case, may be passed directly to subsequent carbon treating station **1** without recombination with the effluent from N separation station **4**.

Example: This example is illustrative of protein biosynthesis by securing production growth of microorganisms using inexpensive C_1 to C_{30} aliphatic hydrocarbon feeds. As is conventional practice, small amounts of salts furnishing potassium, iron, calcium, magnesium, phosphate, sulfate and trace amounts of zinc, manganese, copper and molybdenum were added by the commercial mixture buffered to a pH of about 7.2 in about 1,000 ml. of H_2O.

A microbial agent consisting of a seed of 500 mg. of *Brevibacterium insectiphilium* (ATCC No. 15,528) was introduced. An input of 2 mols of n-hexadecane was added to the mixture through an input valve and the reaction was incubated for a total of about 90 hours at 30°C. for this run.

During the 90 hour reaction period, the incubation mixture was drawn off continuously to a carbon separation station where a bowl type centrifuge separated out the microbial agent.

All of the microbial agent from the separation centrifuge was recombined with some of the effluent product of the centrifuge and passed to a nitrogen treating station physically similar to the carbon treating station, but furnished with an extra cellular source of nitrogen consisting of ammonium sulfate (200 mg./l.). The mixture was agitated and stirred under these proliferating conditions operated so that the whole or average time in the end treating station was about 10 hours. After this time interval was satisfied, the mixture was then passed to a separating station consisting of a centrifuge which removed the nitrogen rich *Brevibacterium insectiphilium* microbial agent.

The microbial agent was split and about 60% was harvested as product and the remaining 40% was recombined with some of the effluent from the centrifuge and returned to the carbon treating station, thus providing nitrogen rich effluent and high nitrogen microbial agent for continuous treatment. In this case the product was in the microbial agent itself growing on the hydrocarbon feed in the carbon and proliferating in the nitrogen treating station.

The cell growth based upon the recovery in the centrifuging from the nitrogen treating station was about 100%. Yield 1,020 mg. *Brevibacterium insectiphilium.*

SPECIALIZED TECHNIQUES

Anaerobic Photosynthetic Fermentations

Microorganisms can oxidize and modify hydrocarbons producing various products and chemicals and increase the cell yield to more protein. However, all of these reactions occur in the presence of air. It has been reported that anaerobic attack by hydrocarbons is possible by sulfate-reducing bacteria which are nonphotosynthetic organisms that derive their energy from sulfate reduction to sulfide.

D.O. Hitzman; U.S. Patent 3,525,671; August 25, 1970; assigned to Phillips Petroleum Company has found that hydrocarbons, preferably normally gaseous hydrocarbons such as natural gas, can be photosynthetically converted by bacterial action into useful products under anaerobic conditions whereby the hydrocarbon serves as the source of carbon.

In the course of the process, the microorganism devours all or a portion of the particular hydrocarbon being utilized as the carbon source. In addition to the growth of the organism, as represented by the cells harvested from the culture medium, a variety of by-products results from the microbial process. Such by-products include vitamins, enzymes, acids, amino acids and alcohols, all of which can be recovered by conventional separation procedures.

In the process the hydrocarbon feed is added to the microorganism or commingled with the photosynthetic microorganism in a growth zone free of air. The microorganisms utilize their intake materials not only for the multiplication of the microorganism cells in number but also for the increase in size of the cells themselves.

The microorganisms preferably employed in the process are those of the family Athiorhodaceae genus Rhodopseudomonas and particularly the species capsulata, spheroides, palustris and gelatinoa.

Example 1: A series of anaerobic modifications using photosynthetic cultures were carried out utilizing a mineral media yeast extract medium containing the following composition:

$(NH_4)_2SO_4$	0.5 g./l.
K_2HPO_4	0.5 g./l.
$MgSO_4 \cdot 7H_2O$	1.0 g./l.
Yeast extract	0.1 g./l.
Burris trace mineral mix	1 cc
Tap water	1,000 cc

The culture medium, inoculated with the microorganism, was maintained at room temperature and at atmospheric conditions with fluorescent lighting provided to the growth zone having a 100% methane atmosphere. Harvesting the culture after 7 days of incubation gave the following results.

Run No.	Microorganism	Dry Weight of Cells, g./l. Charge	Recovered	Growth
1	*Rhodopseudomonas gelatinosa*	0.05	0.55	0.50
2	*Rhodopseudomonas gelatinosa*	0.1	0.95	0.85

The above data demonstrate the cultivation of *Rhodopseudomonas gelatinosa* is achieved under anaerobic conditions utilizing methane as the carbon source.

Example 2: Another series of anaerobic cultivations were carried out in a manner similar to that of Example 1 wherein cultures of the indicated microorganisms were grown in a hexadecane yeast extract mineral medium with a nitrogen atmosphere and with fluorescent lighting. Each of the cultures was carried out at room temperature and pressure. Harvesting the culture after 7 days of incubation gave the following results.

Run No.	Microorganism	ATCC No.	Dry Weight of Cells, g./l. Charge	Recovered	Growth
1	*Rhodopseudomonas capsulata*	11166	0.1	1.3	1.2
2	*Rhodopseudomonas spheroides*	11167	0.15	0.7	0.55
3	*Rhodopseudomonas palustris*	11168	0.1	1.55	1.45
4	*Rhodopseudomonas gelatinosa*	11169	0.05	0.9	0.85

The above data indicate that cultivation of microorganisms under anaerobic conditions is possible utilizing a hydrocarbon such as hexadecane as the source of carbon.

Maturation Treatment at Low Temperatures

W.W. Leathen; U.S. Patent 3,620,927; November 16, 1971; assigned to Gulf Research & Development Company has described a process wherein a hydrocarbon-contaminated microorganism is subjected to a maturation stage during which the hydrocarbon contaminant is greatly reduced or eliminated while simultaneously increasing the yield of product, increasing the protein content of the product and decreasing the amount of ash which is normally formed if the fermentation is continued under normal active fermenting conditions.

In the maturation stage, the hydrocarbon-contaminated microorganism is contacted with a gas containing free oxygen in the presence of an aqueous nutrient medium in the absence of added hydrocarbon at a temperature below the temperature normally employed during the active fermentation stage of the process. Maturation of the microorganism can be conducted in the fermentation vessel or, if desired, the microorganism containing the hydrocarbon contaminant can be separated from the fermentation mass and thereafter contacted in a separate vessel with a gas containing free oxygen in the presence of aqueous nutrient medium.

A microorganism which is substantially free from a hydrocarbon contaminant can be obtained in a process which comprises the following:

(1) cultivating a hydrocarbon-consuming microorganism at a temperature of about 25° to 40°C. in the presence of an aqueous nutrient medium, a hydrocarbon and a gas containing free oxygen;

(2) recovering a microorganism containing a small amount of hydrocarbon as a contaminant;

(3) contacting the hydrocarbon-contaminated microorganism at a temperature below that which supports active fermentation of the microorganism with an aqueous nutrient medium and a gas contaning free oxygen in the absence of added hydrocarbon.

There is thus obtained an increased yield of microorganism substantially free from contaminating hydrocarbon.

The actual temperature at which maturation is conducted in the process depends somewhat upon the particular microorganism being cultured and upon the nutrient medium employed. If the microorganism is actively cultivated at about 25°C. the maturation should be effected at a temperature below 25°C. and preferably at a temperature within the range of about 10° to 20°C. If the microorganism is actively cultivated at about 35°C., the maturation should be effected at a temperature below 35°C. It is preferred to effect maturation at a temperature which is 5° to 20°C. below the active fermentation temperature.

According to the process, active fermentation is interrupted at the end of logarithmic growth and subjected to maturation treatment at a temperature below that which supports active fermentation. During the maturation treatment, the hydrocarbon is greatly reduced or eliminated while the yield of product is increased, the protein content of the product is increased and the amount of ash which is normally formed if fermentation is not discontinued at this point is decreased. Maturation as disclosed hereinabove is effected at a temperature below the temperature at which active fermentation is conducted.

Example: 7 l. of an aqueous mineral nutrient medium are introduced into a 14 l. Pyrex glass fermentor jar equipped with a stainless steel head assembly. The head assembly contains ports for the addition of nutrients and removal of samples, an agitator shaft, an air sparger line, baffles and a thermometer well. The nutrient medium introduced into the fermentor has the following composition

Monopotassium hydrogen phosphate, KH_2PO_4	2.0 g.
Magnesium sulfate, $MgSO_4{\cdot}7H_2O$	1.0 g.
Sodium chloride, NaCl	0.2 g.
Calcium chloride, $CaCl_2{\cdot}2H_2O$	0.2 g.
D-biotin	5 μg.
Ammonium sulfate, $(NH_4)_2SO_4$	0.66 g.
Boric acid, H_3BO_3	1,000 μg.
Copper sulfate, $CuSO_4{\cdot}5H_2O$	80 μg.
Potassium iodide, KI	200 μg.
Ferric chloride, $FeCl_3{\cdot}6H_2O$	400 μg.
Zinc sulfate, $ZnSO_4{\cdot}7H_2O$	800 μg.
Distilled water	1,000 ml.

The fermentor is placed in a water bath which is adjusted to maintain the fermentation medium at 29° ± 1°C. The fermentor is equipped with impellers connected to a drive mechanism capable of rotating the impellers at rates up to 1,000 rpm. The air which is introduced into the fermentor through the sparger is passed through a stainless steel filter tube packed with Pyrex glass wool to remove any particles of dust as well as any airborne microorganisms.

The fermentor which contains the 7 liters of nutrient medium is then inoculated with 25 g. of wet (73% moisture) yeast of the strain *Candida tropicalis* (CS-8-17). The cellular density of the yeast in the fermentor is thus about 6.75 g. of dry matter per 7 l. or about 1 g./l.

The strain of *Candida tropicalis* (CS-8-17) used in this example is retained as stock culture in a stationary condition at 25°C. on a medium consisting of 97% by volume of a commercially available yeast nitrogen base medium (Difco) and 3% by volume of hydrocarbon mixture comprising 93.7% by weight of C_9 to C_{13} normal paraffins, the normal paraffin distribution being substantially as follows:

n-Paraffin	Weight Percent
C_9	Trace
C_{10}	6.8
C_{11}	35.2
C_{12}	32.1
C_{13}	25.8

Active fermentation is conducted for 15 hours at 29°±1°C. with an aeration rate of about 5 liters per minute at impeller speeds of 500 to 900 rpm. During this 15 hours period a total of 88.7 ml. of the above described C_9 to C_{13} n-paraffin mixture are added in about 6 ml. increments at hourly intervals. During the first three hours of active fermentation, a total of 4.3 ml. of an aqueous solution of ammonium sulfate containing 47.88 mg. of nitrogen per ml. is added in about 1.5 ml. increments.

At hourly intervals during the next 12 hours of active fermentation, a total of 55 ml. of ammonium hydroxide containing about 50 mg. of nitrogen per ml. is added in about 4.5 ml. increments. The total amount of nitrogen added to the fermentor during the 15 hour period of active fermentation is about 3 g. The pH of the fermentation mass is maintained at about 3 during the active fermentation by the addition of the ammonium hydroxide.

At the end of the 15 hour active fermentation period the cellular density is about 7.2 g. of dry matter per liter, the protein content about 40.6% by weight, the ash about 12.7% by weight and the residual hydrocarbon about 0.03 ml. per 100 ml. of brew.

At the end of the active fermentation period, the fermentation mass (brew) without being removed from the fermentor is subjected to a 12 hour maturation period during which neither hydrocarbon nor nitrogen is added. The hydrocarbon remaining in the brew at the start of the maturation period comprises about 0.03 ml. of hydrocarbon per 100 ml. of brew. During the maturation period, the temperature of the brew is reduced below that which will support active fermentation, i.e., below about 25°C. In this example, the temperature during maturation is maintained at about 10°C. with an aeration rate of about 0.5 liter per minute at an impeller speed of about 450 rpm to control foaming. A reduced aeration rate is employed to avoid excessive foaming. A reduced impeller speed is employed to avoid possible disintegration of the cellular material.

At the end of the 12 hour maturation period, only a trace of hydrocarbon remains in the brew, i.e., less than about 0.01 ml. of hydrocarbon per 100 ml. of brew. Yeast cells are harvested by centrifuging the brew to recover a yeast paste. The yeast paste is washed with water and centrifuged (usually about three times) until the desired purity is obtained. The wet yeast cells are then dried at a temperature less than about 100°C., i.e., about 60°C. to preserve the quality of the product. The moisture content of the dried product (*Candida tropicalis,* CS-8-A17) is less than 10% by weight of the product. The yield of product is 7.7 g./l., the protein content of the product is 43.8% by weight and the ash content is 13.8% by weight.

Starvation Phase Effecting a Low Lipid Content

J.A. Filosa; U.S. Patent 3,264,196; August 2, 1966; assigned to The British Petroleum Company Limited, England has described a process for the removal of absorbed hydrocarbon and lipids from the yeast cells which have been grown on hydrocarbons.

Example: 40 liters of an aqueous mineral nutrient medium having the composition given below were introduced in a stainless steel fermentor having an effective capacity of 60 liters.

In order to keep the temperature in the fermentor constant at 30°C., water was circulated in an annulus constituted by the space between two concentric cylinders, the smaller one being the fermentor itself. The aqueous nutrient medium had the composition:

	Grams
Diammonium phosphate	2
Potassium chloride	1.15
Magnesium sulfate $\cdot 7H_2O$	0.65

(continued)

	Grams
Zinc sulfate	0.17
Manganese sulfate·H_2O	0.045
Ferrous sulfate·$7H_2O$	0.068
Yeast extract	0.025
Tap water	200
Distilled water q.s. 1,000 ml.	

20 liters of a 24 hour inoculum of *Candida lipolytica* on mixed C_9 to C_{16} normal hydrocarbons were then added, such that the cellular density was about 1 g. of dry matter per l. 1.03 l. of heavy gas oil, that is 15 g./l., were then added, this quantity being sufficient to take the cellular density to 2 g./l.

The temperature of the culture was kept at 30°±1°C., pH at 4 and aeration and agitation giving 3 mmols of O_2 per liter of medium per minute. 10 N ammonia was admitted by an automatic pH controller.

When the flow of ammonia reached 20 ml. the addition of gas oil was begun assuming a yield on gas oil of

$$100 \times \frac{\text{dry yeast produced}}{\text{gas oil feed}}$$

of 10% by weight and a cell division time of three hours. This addition was carried out every hour until 200 g./l., i.e., 13.8 l. had been added.

Starting with a cellular density of 2 g./l. at 25 hours at the end of the exponential growth phase, a density of 15 g./l. was obtained. This value remained constant during the stationary phase from 25 to 40 hours. During this period the yeasts were lacking in carbon feedstock, other elements being present in excess. It acted as a phase of starvation.

The yeasts were recovered at 25 hours and at 40 hours by centrifuging and washing at +5°C. They were then dried by lyophilization and analyzed. Their total lipid content (fraction soluble in hexane) was 18% when recovered after 25 hours and 3% when recovered after 40 hours.

Controlling the Introduction of Additional Hydrocarbon Substrate

The concentration of a moderately volatile hydrocarbon substrate in a fermentation medium can be readily determined and that concentration maintained at a predetermined level by monitoring the vapor concentration of the hydrocarbon in the fermentation exhaust gases and controlling the introduction of additional substrate into the fermentation medium based on the amount of the vapor content. This can advantageously be carried out in a continuous manner by automated instrumentation according to the description of *P. Hosler; U.S. Patent 3,586,605; June 22, 1971; assigned to Sun Oil Company.*

The microbiological oxidation of hydrocarbons to form oxidative products thereof has been investigated and reported in numerous recent literature. Not well known or understood is the need for controlling the concentration of the hydrocarbon substrate in a fermentation medium in order to achieve improved yields.

There is the desirability of controlling the rate of addition of the hydrocarbon feed so that the amount thereof in the dispersion in excess of the amount absorbed by the microorganism cells does not exceed the limit of solubility of the hydrocarbon in the aqueous nutrient medium. This thus-desired control is achieved by taking samples of the fermentation medium from time to time, testing them to determine the concentration of hydrocarbon substrate in the aqueous phase of the medium and regulating the introduction of additional feed accordingly.

While the effect of this sampling and control is to increase the yield of oxidative product, nevertheless complex testing procedures involving centrifuging the samples, extracting them with suitable solvents and analyzing the extract by UV spectroscopy required at each sampling interval. Despite these elaborate methods, such steps are only partially successful in achieving the desired control.

Accordingly, it is an object of this process to provide an improved method and apparatus for closely and accurately measuring and controlling the concentration of a hydrocarbon substrate at a predetermined level in a fermentation medium where the hydrocarbon is undergoing microbiological oxidation.

It has been found that the concentration of a moderately volatile hydrocarbon substrate in an aqueous fermentation medium can be readily determined and the concentration closely controlled and maintained at a predetermined level by:

(1) establishing under aerobic fermentation conditions an

aqueous fermentation medium comprising a hydrocarbon-oxidizing microorganism and a predetermined amount of a liquid hydrocarbon substrate dispersed in an aqueous nutrient medium;

(2) sampling the exhaust gas from the fermentation to determine the vapor content of the hydrocarbon in the exhaust gas; and

(3) controlling the introduction of additional hydrocarbon substrate into the fermentation medium in dependence upon the vapor content of the hydrocarbon in the exhaust gas in order to maintain the predetermined concentration.

Figure 4.4 is a schematic view of a fermentor in combination with hydrocarbon vapor analyzer which actuates a feed pump for regulating the rate of introduction of hydrocarbon feedstuff into the fermentor.

In oxidizing alkylbenzenes of the C_7 to C_{10} range, best results are generally obtained when the unconsumed hydrocarbon concentration is maintained at a level of about 5 to 150 ppm in the aqueous nutrient phase. For other types of substrates, and for any given microorganism, this optimum level can conveniently be determined experimentally by varying the hydrocarbon concentration levels and observing both the product yield and rate at which it is produced, as well as monitoring the oxygen and carbon dioxide levels in the exhaust gas of the fermentor. A decrease in carbon dioxide and an increase in oxygen, for example, is indicative of an oversupply of hydrocarbon feed. The reverse is true when the feed supply is inadequate.

Time-consuming and laborious steps can be avoided, and the concentration of hydrocarbon in the fermentation broth controlled much more closely and accurately by directly measuring the vapor content of the hydrocarbon feedstuff in the exhaust gases of the fermentor without having to resort to taking samples of the fermentation broth. The concentration of hydrocarbon in the exhaust gas is directly proportional to the concentration of hydrocarbon in the aqueous fermentation.

There thus exists an equilibrium distribution of the hydrocarbon between the aqueous fermentation and the fermentation exhaust gas which conveniently permits the monitoring of the hydrocarbon

FIGURE 4.4: CONTROLLING HYDROCARBON FERMENTATIONS

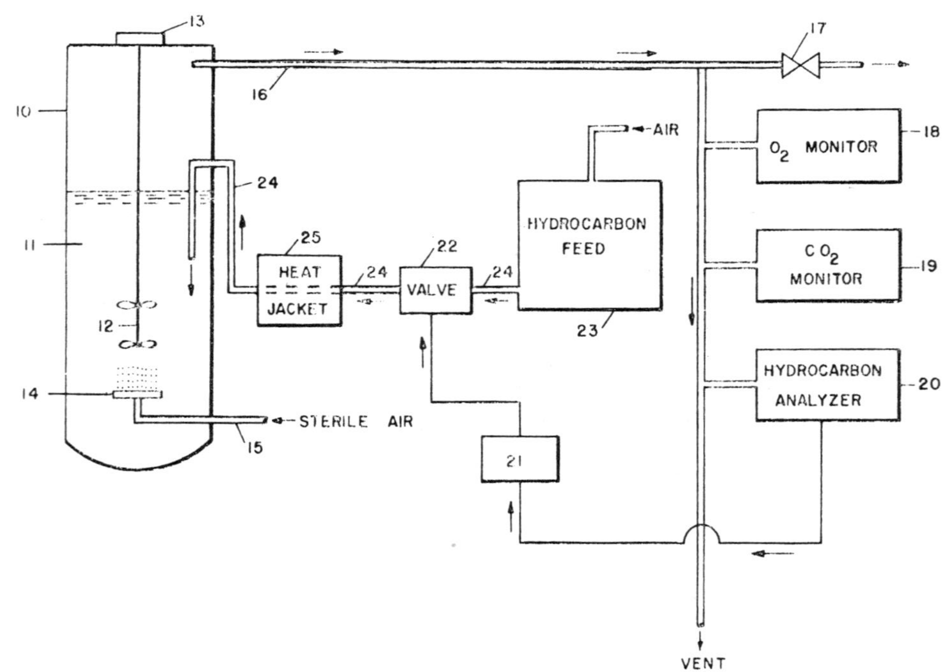

Source: P. Hosler; U.S. Patent 3,568,605; June 22, 1971

vapor content of the exhaust gas by appropriate measuring devices. Preferably, continuous monitoring by automatic instruments is provided where the signal from the exhaust gas analyzer may then be used to determine the hydrocarbon feed rate, desirably by actuating a feed mechanism.

The direct correlation that has been found to exist between the vapor content of the hydrocarbon in the exhaust gas and its concentration in the aqueous phase of the fermentation broth is apparently based on the principle that the partial vapor pressure of any volatile constituent of a solution is equal to the vapor pressure of the pure constituent multiplied by the mol fraction of that constituent in solution. Based on this principle the concentration in the aqueous phase can readily be determined from the vapor content and the rate of feed to the fermentation broth conveniently controlled and maintained by appropriate devices.

The limits of the relationship between the vapor phase content and the aqueous phase content of the hydrocarbon are the solubility of hydrocarbon in water, and the saturation of the exhaust air stream, as influenced by the partial pressure of the hydrocarbon at fermentation temperature. In the case of a fermentation of xylene, for example, where the xylene is fermented at 30°C., the distribution has been found to follow the following relationship:

1 mg. xylene/l. of water = 0.84 mg. xylene/l. of exhaust gas.

The limit of the above relationship is approximately 37 mg. of xylene per liter of exhaust gas, which is the saturation point.

Figure 4.4 illustrates one typical method of operation which is applicable to different combinations of microorganisms and hydrocarbon substrates.

In the figure there is shown fermentor **10** containing culture medium **11** which has been inoculated with a suitable microorganism. The fermentor is fitted with agitator **12** driven by motor **13**. The agitator may be one or more in number depending upon the degree of agitation and aeration desired for any given fermentation. The fermentor is also fitted with an aerator **14** through which sterile air can be bubbled from pipe **15**. Optionally the fermentor may be fitted with a pH measuring device which actuates a container of alkali, in order to maintain a given pH level.

At the top of the fermentor the gases are exhausted with pressure from the pressurized fermentor through pipe **16** fitted with a back-pressure regulating valve **17**. These gases are then passed along pipe **16** to which are attached oxygen monitor **18**, carbon dioxide monitor **19**, and a hydrocarbon analyzer **20**. The remaining unsampled gases are then vented from pipe **16**.

The oxygen and carbon dioxide monitors, which may be of any conventional construction, are employed for the purpose described earlier of measuring the respiratory activity of the cells, which in turn is a measure of their product-forming ability. While monitors **18** and **19** are not essential to the performance of this process, they do simplify the overall control of the fermentation system, and particularly the startup procedures.

The electrical signal from hydrocarbon analyzer **20** is used to activate pneumatic converter **21**, which may be of any conventional design. The regulated air output of this converter is, in turn, used to activate pneumatic valve **22**. The hydrocarbon feed, in a pressurized vessel **23**, is thus supplied through pipe **24** to the fermentor **10** in a proportional mode of operation to maintain a predetermined concentration of hydrocarbon substrate dispersed in the aqueous medium.

If it is desired to maintain asepsis of operation, the hydrocarbon can be sterilized by passing it through a steam-jacketed heat exchanger **25**. The startup of this operation is ordinarily done manually to avoid initial "overshoot" until the overall system has reached equilibrium by responding to the vapor content of the exhaust gas.

As an alternative to proportional control the hydrocarbon feedstuff can be pumped to the fermentor using a simple relay to provide off-on control of a pump attached to vessel **23**, which relay is activated by the output of the hydrocarbon analyzer. This mode of control can be quite effective if care is taken to set the pumping rate only slightly in excess of required feed rate so that excessive over-supply of hydrocarbon is avoided.

The hydrocarbon analyzer may be one of several different types which are commercially available. Thus, for example, a unit which measures ultraviolet light at the particular wavelength of the particular hydrocarbon feedstuff is suitable for purposes of this process, especially if the hydrocarbon feed is an aromatic compound. Alternatively, a total hydrocarbon analyzer with a flame ionization detector may be employed instead, particularly if the hydrocarbon is a straight chain alkane.

The process is applicable to a wide variety of procedures involving the fermentation of hydrocarbon-oxidizing microorganisms in combination with various hydrocarbon substrates to produce many different types of oxygenated hydrocarbon products. Microorganisms which may be employed in these fermentations include bacteria, yeasts, molds and actino mycetes, and particularly such genera as Micrococcus, Corynebacterium, Nocardia, Pseudomonas, Mycobacterium, Streptomyces, Aspergillus, and Acetobacter.

The hydrocarbon substrates include any moderately volatile, normally liquid hydrocarbon having from 5 to 12 carbon atoms, and particularly the aromatic hydrocarbons such as the C_6 to C_{10} benzenoid hydrocarbons. Thus, for example, such hydrocarbons as benzene, toluene, xylene, hexane, octane, decane, dodecane and biphenyl or the like may suitably be employed in this process.

The following examples illustrate applications to three different types of fermentative processes involving, in Examples 1 and 2, the production of three different organic acids, and in Example 3 the growth of a starter culture of cells on benzene, which culture is useful in the fermentation of subsequent hydrocarbon substrates.

Example 1: Nocardia corallina ATCC No. 19,070 was used to prepare dimethyl muconic acid (DMMA) from p-xylene in a 760 liter fermentor of the type shown in Figure 4.4, together with an automatic pH monitoring system. A mineral salt solution of the following composition was used as the aqueous nutrient medium:

	Concentration, g./l. of H_2O
$MgSO_4 \cdot 7H_2O$	0.2
Na_2CO_3	0.1
$CaCl_2 \cdot 2H_2O$	0.01
$MnSO_4 \cdot H_2O$	0.02
$FeSO_4 \cdot 7H_2O$	0.005
Na_2HPO_4	3.0
KH_2PO_4	2.0
Urea	2.0

520 liters of the solution were charged and sterilized and the mixture was inoculated with the organism. The mixture was stirred

vigorously at 30°C. while being aerated at a constant rate of 2 cfm. Normal paraffins were used as a growth substrate and were added at the rate of 100 ml./hr. Also the pH level was automatically held at 6.8 during the run by addition of aqueous caustic soda using a pH controller.

The exhaust gas was monitored by an ultraviolet absorption instrument, and the control apparatus was set to maintain 7 mg. xylene per liter of exhaust gas. At the end of the fermentation at about 50 hours, analysis of the whole broth showed that it contained about 15.4 g. of DMMA per liter.

Example 2: In this run *Nocardia salmonicolor* ATCC No. 19,149 was used to oxidize p-xylene to dihydroxy p-toluic acid (DHPT) and p-toluic acid (PTA). The procedure and apparatus were the same as in Example 1. 34 liters of a mineral salt mixture were sterilized in a 60 liter fermentor provided with aeration and agitation. The mixture was inoculated with the organism and stirred vigorously at 30°C. while being aerated at a constant rate of 9 liters per minute.

Normal paraffins were used as a growth substrate and were added at the rate of 7 ml./hr. At 12 hours the cell mass had increased to approximately 3 g. dry weight/l. At this time the pH was raised from about 7 to pH 8, and the xylene maintained at 20 g./l. of exhaust gas. After 20 hours of the controlled xylene addition the fermentation broth was analyzed and found to contain 4.2 g. of DHPT and 5.3 g. of PTA per liter.

Example 3: The culture *Nocardia salmonicolor* ATCC No. 19,149 was grown on benzene as a growth carbon source to provide quantities of cell mass of the culture. 34 liters of a mineral salt mixture were sterilized in a 60 liter fermentor provided with aeration and agitation. The mixture was inoculated with the organism and stirred vigorously at 30°C. while being aerated at a constant rate of 6 liters per minute. Benzene was added to the system described above in Example 1 to maintain 12 mg. benzene/l. of exhaust gas. After 25 hours of operation growth was determined by centrifugation at 3,600 rpm. A packed cell volume of 3% was obtained.

Reduction of Nucleic Acid Content of Cells

Microorganisms, for example yeasts, grown on hydrocarbon substrates have a high content of nucleic acids, which makes their use in food objectionable. The process of *C. Gatellier and G. Glikmans; U.S. Patent 3,702,283; November 7, 1972; assigned to Institut Francais du Petrole des Carburants et Lubrifiants, France,* obviates this problem by providing a process according to which a microorganism previously grown on a hydrocarbon substrate is contacted with oxygen and a multiplication inhibitor, specifically carbon dioxide, thereby stopping the multiplication while maintaining substantially unchanged the other life functions of the microorganism cells.

It is now possible to achieve a substantial decrease of the nucleic acids content using unwashed cells contained in a reactor while retaining the essential life functions of the cell. It has also been observed that, when using again the conventional operating conditions, i.e., when washing the cells before contacting them with the inhibitor, no reduction of the nucleic acids content was observed.

The inhibitor is an antibiotic, for example, mitomycin C or an actinomycin. The growth may be inhibited with any known antibiotic provided it does not inhibit the breathing activity of the microorganism cells.

Certain antibiotics such as mitomycin C are quite selective, so that there is practically no upper limit for them. In other words, with these antibiotics used in any amount higher than the minimal amount, the growth is stopped but the other life functions are retained.

On the contrary, with less selective antibiotics, for example, chloramphenicol, the amount of antibiotic has to be selected between an upper and a lower limit.

The chemical inhibitor is an amino acid analog, the dose of which has to be determined in the same manner as that used with the antibiotics (minimal and, if necessary, maximal amounts). The following may be used, for example: 5-methyl-tryptophane which is an analog of L-tryptophane; para-fluoro-phenylalanine which is an analog of L-phenylalanine; ethionine which is an analog of L-methionine.

According to another aspect of the process, this inhibitor is an analog of the purine and pyrimidine bases, for example, 8-azaguanine and malonyl-urea or carbon dioxide which may be used in admixture with the oxygen necessary to the metabolism of the microorganisms, at least one part thereof dissolves in the aqueous phase.

There is used, for example, air containing 0.5 to 2% and preferably 0.6 to 1% by volume of CO_2.

Example 1: A *Candida lipolytica* yeast of the *Endomycopsis lipolytica* type is cultivated in a continuous manner, while maintaining growth limitation conditions, the limiting factor being the petroleum cut having a high content of C_{11} to C_{18} normal paraffins (main substrate).

This culture is carried out in a stirred cylindrical reactor of 1 cubic meter capacity; air is blown at a rate of 100 cm. per hour; the pressure being 2.5 atmospheres and the temperature 30°C ± 1°C.

The used hydrocarbon cut is obtained by dewaxing a gas oil cut by means of urea; it consists of 90% normal paraffins and 10% of a mixture of branched paraffins, naphthenes and aromatics. It is introduced into the reaction vessel at a rate of 1 kg. per hour, together with an inorganic solution (about 150 l./hr.) providing 40 g. of phosphorus (P_2O_5), 35 g. of potassium (K_2O) and 1 g. of magnesium (MgO) per hour, these elements being supplied as potassium phosphate and magnesium chloride.

The pH is maintained at 4±0.1 by means of ammonia used at a rate corresponding to 100 g. of nitrogen (N) per hour. The inorganic solution contains the other salts necessary for growth in sufficient amounts.

The effluent from the reactor gives per hour, 1.5 kg. of yeast (dry weight) and 150 liters of an inorganic solution containing 5 g. of phosphorus (P_2O_5) (however it practically no longer contains potassium or magnesium), together with 0.1 kg./hr. of hydrocarbons. Analysis of the raw cells is as follows:

Nitrogen (Kjeldahl) – 9.2%
Nucleic nitrogen (calculated with respect to the Kjeldahl nitrogen) – 10.3%.

The effluent is continuously passed to a second reactor at 30° ± 1°C., the latter being fed with fresh air from a blower. The pH is maintained at 4.5 ±0.2 by means of ammonia. Mitomycin C is present at a constant concentration of 10 micrograms per liter.

The effluent of this second reactor contains 1.6 kg./hr. of yeast (dry weight) and 150 l./hr. of an inorganic solution; paraffins are no longer present. After centrifugation, washing and drying, there are obtained yeasts, the analysis of which gives:

Nitrogen (Kjeldahl) – 10.1%
Nucleic nitrogen (with respect to the nitrogen Kjeldahl) – 5%.

Example 2: Example 1 is repeated except that mitomycin C is substituted with para-fluoro-phenylalanine at a concentration of 2μg. per liter. The resulting yeasts have the following composition:

Nitrogen (Kjeldahl) – 9.3%
Nucleic nitrogen (with respect to the nitrogen Kjeldahl) – 3.5%.

Example 3: Example 1 is repeated except that the effluent from the first reactor is centrifuged to concentrate the cream. The resulting cream gives 1.5 kg. of yeasts and 8 liters of inorganic solution per hour. The cream is conveyed to the second reactor containing mitomycin C, the conditions being those of Example 1. There is thus obtained 1.6 kg. of yeast per hour, the analysis of which gives:

Nitrogen (Kjeldahl) – 9.9%
Nucleic nitrogen (with respect to the nitrogen Kjeldahl) – 6%.

Recovery Techniques

SEPARATION OF CELLS FROM FERMENTATION MEDIA

Controlled Froth Flotation

One serious drawback to the wide acceptance of petroleum biosynthesis technique resides in the fact that it is difficult and very expensive to achieve during biosynthesis sufficiently high concentrations of the bacteria cells produced. The failure to achieve economically satisfactory high concentrations of bacteria cells in the biosynthesis bath leads to high separation and drying costs occasioned by the necessity to centrifuge or otherwise separate the small amounts of bacteria cells from the large amounts of water as well as unconverted hydrocarbons, inorganic salts etc., present in the biosynthesis bath and also of course in the aqueous slurry effluent stream (product stream) thereof.

Due to the low concentration of bacteria cells, e.g., approximately 1% by weight in the aqueous biosynthesis bath, combined with the relatively expensive nature of centrifuging or conducting other suitable separation and drying procedures to concentrate the bacteria cells and remove them from the aqueous media; separation and drying procedures can amount to as much as 60% or higher of the processing costs.

Hence, it will be realized that any substantial improvement in the concentration of bacteria cells prior to separation and drying will effect significant economic savings and reduce the cost of producing high protein foods by bacterial synthesis using hydrocarbon feeds as the relatively inexpensive source of both carbon and hydrogen.

M.B. Perkins and L.E. Furlong; U.S. Patent 3,355,296; November 28, 1967; assigned to Esso Research and Engineering Company have established a marked improvement in the production of high protein-containing foods and food supplements produced from bacteria inoculants grown on petroleum feeds that permits the concentration of bacteria cells (prior to centrifugation or other expensive separation and drying procedures) during biosynthesis to an extent of from 3 to 10 times that attained in typical known fermentation procedures prior to any concentration steps.

It is possible to obtain bacterial cell concentrations of up to and even exceeding 10% by weight during biosynthesis and prior to effecting separation and drying. This constitutes as much as a tenfold or more increase in bacteria cell concentration when compared with the bacteria cell concentrations customarily produced by known biosynthesis procedures.

The bacteria cells are concentrated in the same biosynthesis reactor vessel in which the biosynthesis takes place thus effecting further economies in equipment and space.

In most biosynthesis procedures wherein bacteria are grown on petroleum substrates, there is produced a small amount of frothing. In the past, such frothing has been regarded as highly disadvantageous because it was thought to interfere with biosynthesis, clog apparatus and require special materials and equipment to prevent or at least minimize frothing. Consequently, foaming was substantially reduced or eliminated by use of antifoaming devices and/or agents which, although they did succeed to minimize foaming, tended to restrict growth of the bacteria cells being produced in the biosynthesis bath.

According to this process, foaming is not minimized or avoided but rather established, maintained and controlled to serve a useful purpose. The process employs carefully controlled froth flotation in which the product bacteria cells are withdrawn from the froth zone for concentrating bacteria cells while biosynthesis is still occurring in the lower slurry zone.

Moreover, the foaming is conducted purposely to selectively maximize the concentration of cells in the froth zone compared to the cell concentration in the slurry portion of the biosynthesis bath.

Micrococcus cerificans bacteria cells are naturally surface active and tend to locate in the froth zone of the biosynthesis bath in the form of a stable froth or foam. Several important conditions must be established and maintained during biosynthesis in order to effectively utilize froth flotation during biosynthesis to concentrate bacteria cells grown on normal hydrocarbon feeds.

To begin with the biosynthesis bath, itself, must not inhibit frothing. Thus the biosynthesis bath contains a lower slurry zone and an upper froth zone. Thus use of surface active agents which inhibit or prevent frothing is to be avoided in the biosynthesis bath reactor vessel.

Secondly, the weight ratio of bacteria cells in the froth zone of the biosynthesis bath to the bacteria cells in the slurry zone must be greater than 1.0 to 1, usually exceeds 1.5:1 and preferably exceeds 2.0:1, and the weight percent of bacteria cells in the froth mixture removed from the froth zone should exceed the weight percent of hydrocarbons therein. In other words, it is of primary importance to selectively concentrate the bacteria cells in the froth zone while avoiding concentration of hydrocarbons in the froth zone since this is the zone from which the product stream is withdrawn.

The hydrocarbon feed component constitutes the source of carbon and hydrogen for bacteria cell growth and multiplication. This feed is usually the most expensive feed going to the biosynthesis bath. It has been found that the conversion of hydrocarbons to bacteria is greater in the slurry zone than it is in the froth zone.

Hence, the froth zone needs and makes use of less hydrocarbon than the slurry zone. Moreover, it is important in a commercial biosynthesis procedure to achieve product concentration to a predominance on a weight basis in the product stream while avoiding concentration of nonproduct therein in order to reduce feed costs and product separation costs.

Furthermore, it has been demonstrated that a weight predominance of hydrocarbons over bacteria cells in the froth zone not only constitutes a less efficient utilization of hydrocarbons, but also adversely affects the conversion of hydrocarbons to bacteria in the froth zone resulting in a further lowering of the hydrocarbon to bacteria conversion in the froth zone of the biosynthesis bath.

There should be sufficient hydrocarbon in the froth zone to allow for continued bacteria cell growth and multiplication therein. This is especially important where the biosynthesis is conducted continuously. For this purpose it is usually sufficient to establish a froth zone hydrocarbon content approximately equal to that of the slurry zone hydrocarbon content.

However, the froth zone hydrocarbon content can be slightly below that of the slurry zone due to the slightly lower utilization of the bacteria occurring in the froth zone. In this regard satisfactory continuous froth flotation biosynthesis have been achieved over extended time periods where the weight ratio of froth hydrocarbons to slurry hydrocarbons is as low as 0.8 to 1. When the ratio of froth hydrocarbons to slurry hydrocarbons is well below 0.8 to 1, there is an adverse affect upon bacteria growth and multiplication in the froth zone of the biosynthesis bath due to lack of sufficient hydrocarbon feed therein.

For example, when the ratio of froth hydrocarbons to slurry hydrocarbons is below 0.5 to 1, bacteria growth in the froth zone decreases rapidly. Increasing the concentration of hydrocarbons in the froth zone does not accelerate bacteria growth; nor do high froth hydrocarbon to slurry hydrocarbon ratios appear to hinder growth as long as the froth bacteria exceeds froth hydrocarbons on a weight basis.

In fact, it is not unusual for the weight ratio of froth hydrocarbons to slurry hydrocarbons to be as high as 8 to 1, but in such cases the weight ratio of froth bacteria to froth hydrocarbons is usually from 4 to 6 to 1.

For most froth flotation biosynthesis procedures especially continuous froth flotation biosynthesis, the bacteria cell concentration ratio, viz the weight percent bacteria in the froth in proportion to the weight percent bacteria in the slurry, usually exceeds 1.5 to 1 and preferably exceeds about 2.0 to 1. Ratios exceeding 3 to 1 and 5 to 1 have been achieved, e.g. ranging from 5 to 10:1.

The bacteria cell selectivity in the froth zone, viz the weight ratio of froth bacteria to froth hydrocarbons exceeds 1 to 1. Usually the ratio of froth bacteria to froth hydrocarbons exceeds 10 to 1 and preferably exceeds 20 to 1, e.g., usually ranging from 10 to 20 to 1 and even higher.

It is important to conduct the biosynthesis at a high conversion level in the slurry zone as this has some effect on selective concentration of froth bacteria as compared to froth hydrocarbons in the froth zone of the biosynthesis bath. In order to effect optimum concentration of bacteria cells in the froth zone, the slurry zone conversion of hydrocarbons to bacteria should exceed 90%. Usually, it is desirable to maintain the slurry zone conversion of hydrocarbons to bacteria above 95%, and preferably ranging from 90 to 99%.

In order to insure proper slurry zone conversions of hydrocarbon to bacteria, it is important to feed sufficient oxygen for efficient growth and reproduction. For *Micrococcus cerificans,* the preferred bacteria inoculant, this requires maintaining from 0.5 to 1.5 volumes of air per volume of biosynthesis bath per minute.

Less air volumes are required when oxygen enriched air or straight oxygen are used as an oxygen source. The oxygen can be supplied in any form capable of being assimilated readily by the inoculant microorganism, and oxygen-containing compounds can be used as long as they do not adversely affect bacteria growth and conversion of hydrocarbon feed to bacteria. Conveniently, however, the oxygen is supplied as an oxygen-containing gas, e.g., air, which contains from 19 to 22 weight percent oxygen.

Usually it is preferable to employ air, however oxygen enriched air having more than 22 weight percent oxygen, e.g., enriched air having 70+ weight percent oxygen. 90+ weight percent oxygen, or even pure oxygen can be used and have given good results.

The process will be more clearly understood from a consideration of the accompanying drawings describing a preferred apparatus for continuously carrying out this method. Figure 5.1a is a cross-sectional view of the froth flotation-biosynthesis reactor system, and Figure 5.1b is a cross-sectional view of a froth concentrator used to concentrate the froth mixture issuing as a product stream from the froth flotation-biosynthesis reactor.

Referring to the drawing, wherein like reference numerals indicate like apparatus parts, raw hydrocarbon feed is fed through line **1** to feed purification unit **2** where nonnormal materials and impurities are removed through line **3.** This feed purification is preferably conducted on C_{11} to C_{30} hydrocarbons by 5A molecular sieves followed by desorption and then a cleanup of the desorbed normal hydrocarbons with 13X sieves or silica gel to adsorb remaining impurities.

According to a preferred embodiment, a substantial weight percentage, e.g. 70+ percent of the hydrocarbon feed is comprised of normal

FIGURE 5.1: CONTROLLED FROTH FLOTATION

a.

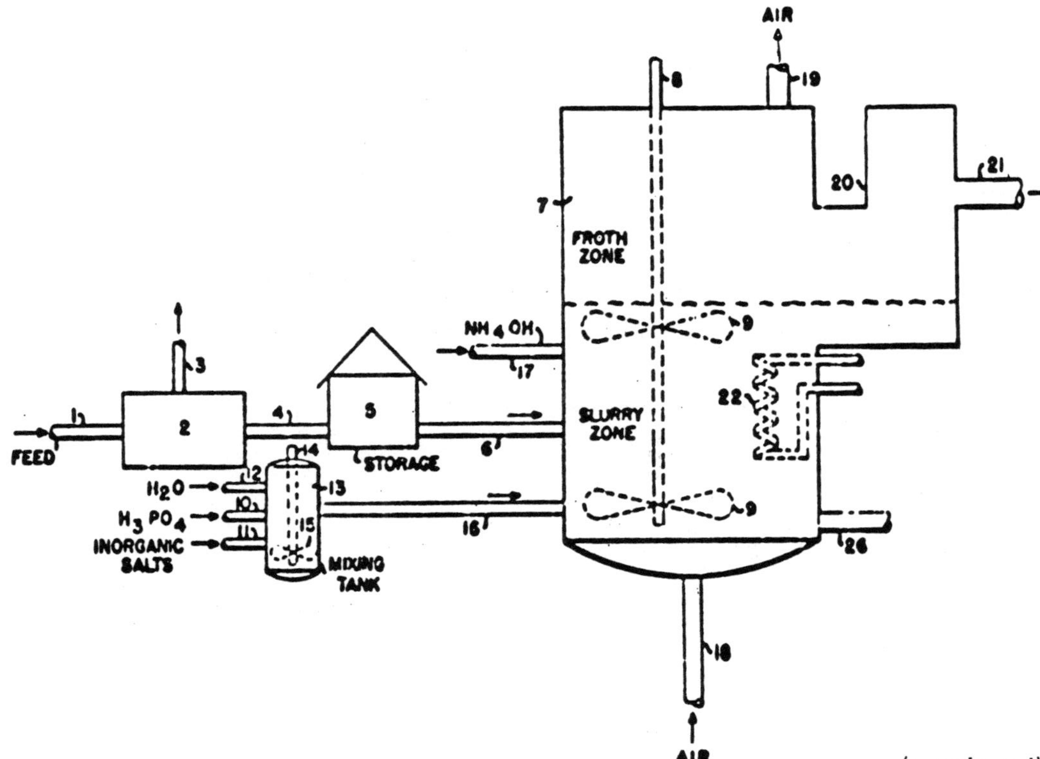

(continued)

FIGURE 5.1: (continued)

b.

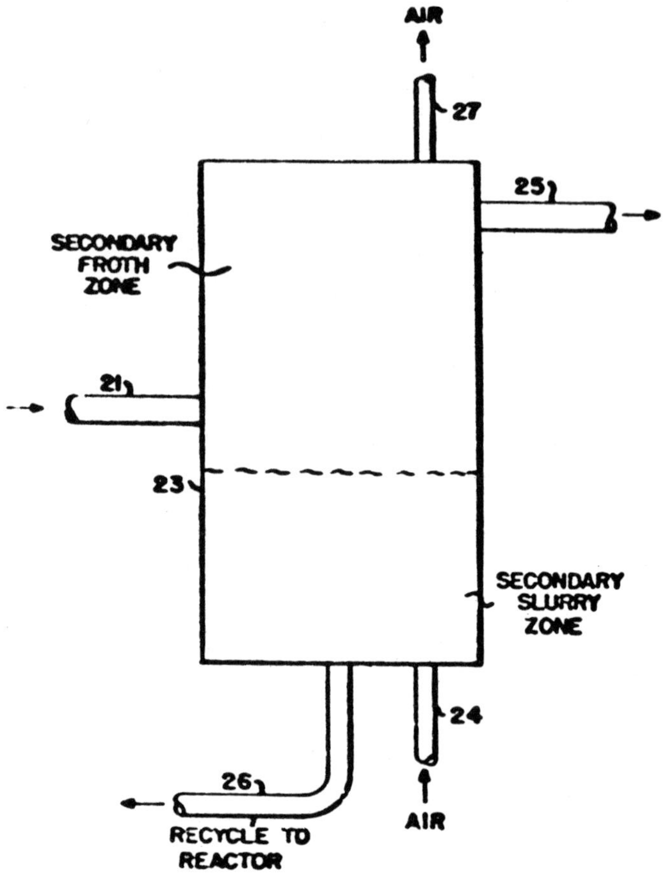

Source: M.B. Perkins and L.E. Furlong; U.S. Patent 3,355,296; November 28, 1967

(straight chain) paraffin hydrocarbons having from 1 to 35 carbon atoms. While the presence of branched, nonaromatic hydrocarbons in amounts up to 30% by weight in the hydrocarbon feed can be tolerated, concentrations in excess of 10 weight percent of nonnormal, nonaromatic hydrocarbons are usually avoided since the bacteria cells employed herein are selective preferentially to normal hydrocarbons, especially n-paraffins. Therefore the use of branched paraffins is preferably avoided. The purified normal hydrocarbons usually contain 90+ weight percent of C_1 to C_{35} normal paraffins and up to 10 weight percent normal olefins.

The purified hydrocarbon feed is passed through line **4** to hydrocarbon storage unit **5**. These normal hydrocarbons are then supplied through line **6** to froth flotation-biosynthesis reactor **7** containing the bacteria inoculant biosynthesis bath which is stirred by stirrers **9** mounted on common shaft **8**. Phosphoric acid is supplied through line **10**; inorganic salts are supplied through line **11** and water is supplied through line **12** to mixing tank **13** equipped with a stirrer **15** mounted on shaft **14**. The mixed salt solution is supplied through line **16** to reactor **7**. Ammonium hydroxide is supplied to the reactor through delivery line **17**.

Air or oxygen enriched air is supplied to the reactor through delivery line **18**. Effluent gases containing unutilized O_2, N_2, CO_2 and water vapor are vented to the atmosphere from the reactor through outlet line **19**. The froth flotation biosynthesis reactor preferably contains a froth flotation removal section **20** which, due to the outward and upward flow of exhaust gases, allows the froth from the upper froth zone to be collected prior to removal through froth removal line **21**.

A cooling coil **22** or other suitable cooling device is present in the lower slurry zone of the biosynthesis bath to aid in maintaining the proper growth temperature for the bacteria. The froth mixture removed from the froth zone of the biosynthesis bath at line **21** contains bacteria cells, unconverted hydrocarbons, and inorganic salts along with water.

The bacteria cells can be separated from the remainder of the froth mixture by conventional centrifuging or settling of the froth mixture directly after collection thereof; however, it is preferable to subject the froth mixture obtained at line **21** to a secondary froth flotation step in froth mixture concentrator **23** (Figure 5.1b) prior to separation of the bacteria from unconverted hydrocarbons, inorganic salts and water.

For this purpose the froth mixture from reactor **7** is carried as a primary product stream from reactor **7** in line **21** to concentrator **23**. A froth breaker (not shown) can be employed in line **21** between reactor **7** and concentrator **23** to break the froth and facilitate delivery thereof to the concentrator, which in effect is a secondary froth flotation cell. Concentrator **23** differs from froth flotation reactor **7** in that it need not serve as both a biosynthesis reactor vessel and a froth flotation cell. In fact very little biosynthesis usually takes place in concentrator **23** compared to reactor **7**. Of course, it is within the purview of this process to run a plural stage biosynthesis wherein concentrator **23** serves as a biosynthesis reactor vessel.

Air or other oxygen rich gas is delivered to concentrator **23** through delivery pipe **24** located in the bottom of the concentrator. The bubbling of air through the froth mixture aids the development of a lower secondary slurry zone and an upper secondary froth zone in concentrator **23**. The concentrated froth mixture is removed as a product stream from the concentrator through froth effluent line **25** for centrifuging and drying. The secondary slurry effluent from the secondary slurry zone is recycled through conduit **26** to reactor **7**. Air and other exhaust gases are vented from concentrator **23** through outlet line **27**.

The bacteria cells in the concentrated froth mixture effluent can then be separated from the remaining unconverted hydrocarbons, inorganic salts and water by conventional settling or centrifuging followed if desired by drying, e.g., in conventional spray driers utilizing hot air to effect drying. Drying is not mandatory however, and the separated bacteria cells can be employed directly as a food supplement without drying.

Instead of using a single concentrator to effect concentration of the bacteria cells in the froth mixture withdrawn from reactor **7**; two, three or even more such concentrators can be arranged in series. The use of a single froth flotation concentrator cell **23** is capable of doubling the concentration of bacteria cells in the froth mixture withdrawn as a product stream from reactor **7**.

The purified hydrocarbon feed usually contains 90+ weight percent of C_1 to C_{35} n-paraffins and up to 10 weight percent of n-olefins. When the inoculant microorganism is *Micrococcus cerificans* a fairly high level of the C_1 to C_{35} n-paraffins are C_{11} to C_{30} n-paraffins, e.g., from about 40 to 99+ weight percent of the C_1 to C_{35} n-paraffins are comprised of C_{11} to C_{30} n-paraffins.

The preferred purified hydrocarbon feed contains 95+ weight percent of C_{11} to C_{30} n-paraffins and up to 5 weight percent normal olefins containing from 11 to 30 carbon atoms. The C_{11} to C_{30}

n-paraffin feeds can be petroleum feeds, e.g., gas oils boiling in the range of 190° to 400°C., and more preferably 190° to 320°C. and containing a substantial amount of C_{11} to C_{30} n-hydrocarbons.

The process extends to the froth flotation biosynthesis of all microorganisms, including bacteria and yeasts, capable of growth on C_1 to C_{35} hydrocarbon feeds including C_1 to C_5 gaseous paraffins, e.g., methane, ethane, propane and gases containing them; C_6 to C_{10} light naphthas, viz, low boiling hydrocarbon oils of the C_nH_{2n+2} series having a boiling point of 95° to 150°C. and petroleum fractions containing them; and C_{11} to C_{30} gas oils boiling 190° to 320°C. and petroleum fractions containing them. Each of the above feeds usually contains normal olefins in varying amounts.

When the microorganisms are grown on methane or other gaseous paraffin feeds, the preferred class of microorganism is Pseudomonadaceae, such as *Pseudomonas methanica.*

When the froth flotation-biosynthesis is performed using a light naphtha feed, the preferred classes of microorganisms are Pseudomonadaceae and Arthrobacter, such as *Pseudomonas fluorescens, Pseudomonas desmolyticum, Pseudomonas aeruginosa* and *Arthrobacter globiforme.*

In the preferred embodiment, the froth flotation biosynthesis is conducted using a bacteria inoculant, especially, gram negative coccus bacteria; but yeasts can be employed, e.g., *Torulopis magnolia* and *Candida albicans.*

While any aerobic, bacteria cells capable of assimilating normal C_1 to C_{30} paraffinic hydrocarbons can be employed, preferred bacteria are as follows: *Micrococcus cerificans, Nocardia opaca* and *Nocardia rubra.* Especially preferred is *Micrococcus cerificans.*

A small extent of mechanical agitation in the form of stirring of the biosynthesis bath is helpful in order to insure proper contact of the bacteria cells with the hydrocarbon, and aqueous growth medium containing the necessary mineral nutrients for bacterial growth and multiplication. Also agitation of the biosynthesis bath helps to cut down on hydrocarbon concentration in the froth zone.

However, the agitation should not be so vigorous as to interfere with the froth flotation by breaking the froth. In fact, excellent bacteria cell concentrations of grown cells have been achieved in the froth zone without stirring. The moderate agitation caused by bubbling air in the form of fine bubbles uniformly throughout the slurry zone provides sufficient agitation for froth flotation.

Excessively high air introduction rates into the biosynthesis bath tended to destroy the froth due to turbulence in the froth itself. However, if insufficient air is bubbled into the biosynthesis bath there results insufficient aeration to insure proper bacteria growth in the slurry zone and froth flotation of bacteria cells.

With regard to the rate of air introduction into the biosynthesis bath, it is important to deliver the air into the biosynthesis bath in a manner sufficient to insure an intermediate or quasilaminar flow of air upwards through the slurry zone to establish proper froth flotation for overhead product removal. For this purpose, the air delivery rate should range from 10 to 100 cubic centimeters/minute/square centimeter of reactor cross section, usually ranges from 15 to 80 cc/min./cm.^{2} and preferably ranges from 20 to 40 cubic centimeters/minute/square centimeter.

Example 1: Micrococcus cerificans were grown on a C_3 to C_{19} n-paraffin hydrocarbon containing no more than 40 ppm aromatics. This feed mixture was obtained by passing San Joaquin virgin distillate (450° to 550°F. vapor temperature) through a 5A molecular sieve and then through a 13X molecular sieve. Chromatographic analysis of a typical feed hydrocarbon showed the following concentrations.

Carbon Number	Weight Percent
8	0.1
9	0.4
10	1.3
11	3.1
12	5.5
13	9.9
14	18.7
15	25.7
16	22.3
17	10.9
18	2.0
19	0.1

The composition of the inorganic salts solution was as follows.

Mineral Salt	Feed Level, wt.
$CaCl_2$	0.010
$FeSO_4 \cdot 7H_2O$	0.0040
KCl	0.050
$MgSO_4$	0.020
$MnSO_4$	0.0040
Na_2SO_4	0.050
NH_4Cl	0.10
85% H_3PO_4	0.15
Water (Elizabethtown, N.J., supply)	Remainder

Growth conditions were established prior to feeding the slurry to the batch flotation tank. The air rate was set at 22 cc/min./cm.^{2} of reactor cross-sectional area. A C_8 to C_{19} hydrocarbon feed rate of 20 g./hr., a mineral salts feed rate of 1,960 g./hr. and a 7.0 N ammonium hydroxide feed rate of 30 g./hr. was set. A pH of 7.0 was controlled by using NH_4OH addition. The temperature was maintained at 34°±1°C. and agitation was continued at 1,000 rpm.

Four liters of approximately 1.0% by weight (dry basis) solids aqueous slurry produced as outlined above were transferred to a froth-flotation device.

The vessel, having a volume of 7,500 cc, was charged with an aqueous slurry containing approximately 1.0 weight percent *Micrococcus cerificans.* Sufficient air was bubbled through the aqueous slurry to produce a heavy froth zone. No stirring was used after transfer of the slurry to the froth flotation device, nor was any other form of agitation employed except that caused by the air. The experiment was conducted at ambient temperatures, viz, 20° to 25°C. The pH was maintained essentially neutral, viz, pH is approximately 7, throughout the run.

Between 5 and 10 minutes after frothing started, representative samples were taken from the froth and slurry. Reaction conditions in this case were: a 75 cc/min/cm.^{2} of reactor cross section air rate and a 0.95 weight percent solids slurry concentration on a dry basis at the start. There was no agitation of the reaction mixture.The resultant concentration ratio of froth concentration to slurry concentration for the cells is 2.24; for the hydrocarbon, 1.41. These samples were analyzed for bacteria cell and hydrocarbon concentration. The bacteria were concentrated 2.24 to 1.0 in the froth as compared to the slurry, thus demonstrating the feasibility of froth flotation to yield bacteria concentrations substantially exceeding those obtainable in conventional biosynthesis baths (where the cell concentrations average approximately 1 weight percent).

Example 2: Experiments were carried out as in Example 1 with the effect of using two stages of flotation being studied. Experiment conditions in this case were agitation in Stage 1 of 1,500 rpm; in Stage 2, 0. The additive used here was 0.2 weight percent of NaCl. The results are shown in the table below.

		Concentration Ratios		
Stage	Cells in Slurry, wt. %	Froth Cells: Slurry Cells	Froth Hydrocarbon: Slurry Hydrocarbon	Air Rate, cc/min./cm.^{2}
1	1.47	2.1	1.9	36
2	3.06	2.0	1.0	16
Product	6.12	-	-	-

The results show how a froth-flotation biosynthesis in combination with a second stage flotation step and 0.2% NaCl can increase the bacteria cell concentration in the product stream. The weight percent of cells in the aqueous slurry of the froth flotation biosynthesis bath was 1.47% compared to 6.12% in the froth coming from second stage for an overall cell concentration ratio of 4.2:1. At the same time the hydrocarbon is concentrated only 1.9:1 thus conserving the hydrocarbon feed in the growth reactor.

Use of Microfine Air Bubbles

In the purification and recovery of products accumulated from fermentation liquors, it is necessary to carry out the removal of the remaining hydrocarbons as well as the separation of the resultant cells on an industrial scale. In order to solve this problem, investigations have been made concerning the fundamental properties of the hydrocarbon fermentation liquor.

As a result of these studies it has been found that all of the fermentation assimilating microorganisms have a property of an extremely strong affinity for the hydrocarbons used in the fermentation of the raw material.

By effectively utilizing this property, *A. Furuya, Y. Aoki and*

M. Takayanagi; U.S. Patent 3,630,843; December 28, 1971; assigned to Kyowa Hakko Kogyo, Ltd., Japan have developed a process for separating and recovering microorganism cells along with hydrocarbons from a water phase.

A much improved process for separating the hydrocarbons remaining after fermentation and the microorganisms cells from the fermentation liquor may be obtained by utilizing the remaining hydrocarbon present in the fermentation liquor after completion of the fermentation or by adding after fermentation additional hydrocarbon of the same type as used in the fermentation process.

Separation is effected by generating microfine air bubbles in the lower portion of the fermentation liquor causing the hydrocarbon and microorganism cells to rise to the upper portion of fermentation liquor where separation of the fermentation liquor into the hydrocarbon microorganism cell phase and the water phase can be readily effected.

In general, in a culture liquor utilizing hydrocarbons as a raw material, there are many cases where the raw material hydrocarbon remains or a cell component is broken. It is extremely difficult, as in the case of the existing saccharine fermentation liquors, to remove cells by a centrifugal separation or to remove cells by filtration with a filtering material.

Further, the broken cell component and the remaining hydrocarbons are emulsified during the fermentation and this makes it difficult to separate the constituents centrifugally or by filtering.

It has been recognized that various hydrocarbon assimilating microorganisms have a property of sedimentation, that is the ability to become suspended in water. But when these microorganisms are suspended in a mixed solution of assimilable hydrocarbon and water and after stirring are permitted to sit, the hydrocarbon becomes separated in an upper phase while the water separates to the lower phase. In such a situation, the microorganism cells are always contained in the hydrocarbon phase, that is the upper phase.

Further, by employing this property, the particular cells and the hydrocarbons can be simultaneously removed and separated in the manner of separating the hydrocarbons from the water phase of a culture liquor in which hydrocarbons remained at the completion of the fermentation or in the case where no hydrocarbons remained in the fermentation liquor, by adding the same hydrocarbon utilized as the raw material.

Furthermore, without employing a special apparatus, the separation can be effectively obtained by generating microfine air bubbles in the fermentation liquor containing the hydrocarbons. By utilizing the remaining hydrocarbons in the liquor or by adding the same hydrocarbon after the completion of the fermentation, and by introducing water containing dissolved air into the lower portion of the fermentation liquor under pressure, the dissolved air from the lower portion of the tank rises in the form of microfine bubbles due to the pressure difference which exists.

This effects the simultaneous floating-up of the hydrocarbons and the microorganism cells and thus the separation of the materials from the water phase section of the fermentation liquor in the lower portion of the liquor.

The hydrocarbons and the microorganism cells which have been floated to the upper portion of the fermentation liquor can be reslurried with water and again floated up and separated by means of the microfine air bubbles, thereby effecting separation and recovery of the hydrocarbons and microorganism cells in the fermentation liquor up to an amount of about 99% or more.

In addition, in the case of applying the process industrially, it is possible to continuously remove the floating mixture of hydrocarbons and the microorganism cell from the upper portion of the liquor and a clear liquor from the lower portion while continuously supplying a fermentation completed liquor and continuously generating air bubbles in the fermentation liquor by dissolving air or other gases therein under an increased pressure in the lower portion of the tank.

The phenomenon of such a strain separation by means of the float-up system is not observed in the case of using a carbon source other than hydrocarbons, that is a carbon source such as for example, a carbohydrate such as starch, glucose and the like. The effect thereof is a peculiar phenomenon observed specifically in a hydrocarbon fermentation liquor.

Example 1: To 100 liters of a fermentation liquor containing 1,600 g. of L-glutamic acid and 4.3% of remaining n-paraffin obtained by culturing the L-glutamic acid producing strain, *Arthrobacter paraffineus,* ATCC 21167 at a pH of 7.2 for 60 hours utilizing 10% (v./v.) of n-paraffin as a main carbon source is added 20 liters of water containing air dissolved therein under a pressure of 2 kg./cm.^{2}. The water is jetted through the lower portion of the tank.

As a result, the liquor is distinctly separated into two phases, a water phase section in the lower portion of the liquor and a section of mixed n-paraffin and microorganism cells in the upper portion. The water phase section from the lower portion of the tank is withdrawn. 100 liters of water is again added to the upper phase and 20 liters of water containing air dissolved therein under a pressure of 2 kg./cm.^{2} is jetted into the lower portion of the tank. The float-up separation is again carried out and the water phase section of the lower portion is separated. The L–glutamic acid recovered in the water phase section is 1,585 g. The recovering percentage is about 99.1% and the recovery percentage of the separated cells recovered together with the n-paraffin is 96.2%.

Example 2: For the purpose of separating cells from 20 liters of a fermentation liquor containing 0.4% of remaining kerosene obtained by culturing *Arthrobacter roseoparaffineus* ATCC 15584 at 30°C. for 4 days utilizing 4.5% (v./v.) of kerosene as the main carbon source, an additional 0.6 liter of kerosene is added to the liquor with stirring at the time of completion of the fermentation. Then 5 liters of water containing air dissolved therein are added to the lower portion of the fermentation liquor under a pressure of 1.5 kg./cm.^{2}.

The liquor is then separated on standing into two phases, a transparent water phase at the lower portion and a phase of kerosene and microorganism cells in the upper portion. After taking out the transparent section from the lower portion of the tank, the cells are recovered together with the kerosene. The percentage recovery of cells is 95%. It should be noted that the separation effect can be increased by the presence of an aggregation agent, and the like in the fermentation liquor.

Addition of Immiscible Organic Liquid

R. Bernheimer and M.R. Friedman; U.S. Patent 3,530,039; Sept. 22, 1970; assigned to Esso Research and Engineering Company have described a process for facilitating the recovery of microorganisms from an aqueous medium.

In particular, this process comprises contacting an aqueous medium containing microbial cells with an immiscible organic liquid to form an aqueous phase and an organic phase; and recovering the microbial cells from the organic phase.

This improved method for harvesting microorganisms from an aqueous medium, comprises, mixing an immiscible, volatile, chlorinated organic liquid with the aqueous medium at a temperature of between 20° and 55°C., and recovering microorganisms from the resulting organic phase.

More particularly, this process, comprises (a) cultivating a microorganism in an aqueous growth medium comprising an inoculant of microorganism cells, a source of carbon, hydrogen and oxygen, and essential cell nutrients: (b) withdrawing a portion of aqueous growth medium containing microorganism cells: (c) mixing an immiscible, organic liquid to form an aqueous phase and an organic phase: and, (d) recovering microorganism cells from the organic phase.

An immiscible organic liquid is mixed or intimately contacted with the aqueous effluent from the fermentation reactor before separation of the product cells. As a result of this mixing, the microbial cells agglomerate and appear to form a stabilized precipitate in the nature of an emulsion within the organic liquid.

The exact manner in which the agglomeration proceeds is, however, not certain. Since the organic liquid is immiscible with the aqueous phase, the mixture easily separates and settles into an aqueous phase and an organic phase. As a result, the two phases can be easily separated, e.g., by centrifugation or decantation, and the product microbial cells recovered from the organic phase by evaporation of the organic solvent and drying of the cells.

The organic liquid employed in the process is, in general, an immiscible organic liquid which preferentially wets the microorganism cells. In addition, it is preferred, although not essential, that the organic liquid have a density greater than the aqueous phase with which it is mixed, i.e., it has a density (D_{H_2O}) of greater than 1.0 and that it be easily volatilized.

These additional characteristics allow for greater ease in separating the organic phase and the aqueous phase and in drying the product microbial cells. An organic liquid is considered to be volatile if its heat of vaporization is below that of water.

While any immiscible organic liquid which preferentially wets the microorganism cells can be employed in the process, there are some which are especially suitable.

They include, halogenated C_1 to C_4 aliphatics (paraffins and olefins), preferably polyhalogenated C_1 to C_4 aliphatics, carbon disulfide and toluene. Specific examples of these halogenated paraffins include: chloroform, methylene chloride, carbon tetrachloride, methylene bromide, 1-bromo-2-chloroethane, etc.

Example: Run 1 – A small glass column with a feed port in the center and take-offs at the top and bottom was charged with 30 cc of the product stream aqueous slurry from a biosynthesis reactor. Ten cc of methylene chloride (dichloromethane) were then added to the glass column and thoroughly mixed with the aqueous slurry.

Admixing temperature was 70°F. Agglomeration of the microorganism cells in the methylene chloride occurred spontaneously upon introduction of the methylene chloride. The admixture was then permitted to settle for 30 minutes. Most of the aqueous phase remained as supernatant. The resulting organic phase was separated from the aqueous phase and the microorganism cells recovered by stripping to remove methylene chloride and occluded water.

Substantially all of the microorganism cells present in the original aqueous slurry were recovered. About 40% of the original aqueous slurry was present in the agglomerated product.

Run 2 – The procedure of Run 1 was repeated except that 30 cc of methylene chloride were employed. Substantially all of the microorganism cells were removed from the aqueous slurry and about 42% of the original aqueous slurry was present in the agglomerated product.

The procedure of Run 1 was repeated except that 90 cc of methylene chloride were employed. Substantially all of the cells were removed from the aqueous slurry and about 50% of the original aqueous slurry was present in the agglomerated product.

Application of Surface Active Agents

A process has been described by *A. Champagnat; U.S. Patent 3,186,922; June 1, 1965; assigned to The British Petroleum Co., Ltd., England* for the recovery of yeasts from an emulsion of yeasts and hydrocarbons in an aqueous mineral medium.

At times when such emulsions are centrifuged using the normal technique which is employed in the manufacture of yeasts from carbohydrates, there is obtained a clear aqueous phase of an aqueous mineral medium.

However, one also obtains a more or less viscous emulsion which contains the yeasts, all or part of the hydrocarbons and a certain quantity of aqueous mineral medium.

The recovery of yeast grown in the presence of hydrocarbons is effected by adding to a mixture of yeast, aqueous mineral medium and hydrocarbon, at least 50 parts per million, and preferably 0.02 to 0.5%, by weight of a surface active agent and thereafter centrifuging, whereby there is obtained a paste comprising yeast and aqueous mineral medium. As the surface active agent, there may be used cationic surface active agents such as stearyl-trimethyl ammonium chloride, nonionic surface active agents, for example the condensates of oleic acid and ethylene oxide, or anionic surface active agents, for example sodium alkyl sulfates.

Preferably, after adding the surface active agent, the mixture is subjected to vigorous mixing and without a further period of growth of the yeast, is subjected to centrifuging.

If the centrifuge is adapted for the separation of two phases only, the centrifuging is controlled to obtain on the one hand the mixture of the two phases, liquid aqueous mineral medium and hydrocarbons, and on the other hand the pasty phase of yeasts in a certain proportion of aqueous mineral medium. In this case the separation of the hydrocarbons and of the aqueous mineral medium is made subsequently either by decantation or, if necessary, by centrifuging.

The yeast cells obtained in the pasty phase are usually still coated by a certain quantity of hydrocarbons fixed principally by adsorption.

The yeast paste which is obtained is diluted with an aqueous medium which may be water alone, or water containing surface active agent and/or mineral constituents, and is again centrifuged.

Suitably, the treatment will consist of several stages of dilution and centrifuging; in the initial stages the yeast is diluted with an aqueous medium containing surface active agent and, in the final stages, the yeast is diluted with an aqueous medium free of surface active agents. In the course of the initial stages the yeast is freed from hydrocarbons; in the final stages the yeast is freed from surface active agents.

If the pH of the mixture to be treated is below 7, it should be raised to a value between 7 and 9; this has been found to facilitate the separation of the phases of the emulsion. This pH correction should be made before the first dilution stage of the process.

Separation of the yeast paste is preferably carried out at a slightly elevated temperature, usually above 30°C. but not exceeding 99°C. The higher the temperature the more one can reduce the dose of surface active agent, but one should not reach a temperature which could alter the cells of the yeast. A temperature of about 60°C. is preferred.

It is advantageous not to waste the aqueous washing liquids which contain the surface active agent; these aqueous liquids may be recycled and used several times. The hydrocarbons that they contain may be continuously separated by decantation or centrifuging.

Example 1: The emulsion from which it is desired to extract the yeasts is constituted by cells of *Saccharomyces cerevisiae* and a paraffinic gas oil dispersed in small particles in a continuous aqueous phase of a mineral medium having the following composition:

Monopotassium phosphate	7 g.
Magnesium sulfate	0.2 g.
Sodium chloride	0.1 g.
Ammonium chloride	2.5 g.
Drinking water	100 ml.
Distilled water, q.s.	1,000 ml.
pH	4

Experiments were carried out on samples taken from the original emulsion.

(1) Intermittent centrifuging of the emulsion as such gave a pasty phase of yeast cells impregnated with hydrocarbons with aqueous medium; a cloudy phase of aqueous mineral medium; a very stable emulsion phase of yeast, mineral medium and hydrocarbons; and a wet hydrocarbon phase.

(2) There was added to the original emulsion one part per 1,000 by weight with reference to the aqueous phase of the nonionic surface active agent Cemulsol, (a condensate of oleic acid and ethylene oxide) and the mixture vigorously mixed. Centrifuging gave a pasty phase of yeast cells impregnated with hydrocarbons with aqueous medium; a clear phase of aqueous mineral medium; and a wet hydrocarbon phase. Doses of surface active agent below 1 part per 1,000 were less efficient. These two experiments were carried out at a pH of 4.

(3) The pH of the original emulsion was adjusted to 8 and to this was added only 0.5 part per 1,000 with reference to the aqueous phase of the same surface active agent. Centrifuging gave the same results as (2). (1), (2), and (3) were carried out at a temperature of 30°C.

(4) (3) was repeated at a temperature of 60°C. The separation by

centrifuging was clearly better than in (3) since the pasty phase of yeast cells was less rich in hydrocarbons which entered for the greater part into the wet hydrocarbon phase.

(5) The pasty phase of yeast cells impregnated with hydrocarbons and with aqueous mineral medium from (4) was washed with water at 90°C. and centrifuged. There was obtained: a pasty phase of yeast cells still containing hydrocarbons with an aqueous phase; and an aqueous phase containing hydrocarbons. Four successive operations of washing and centrifuging at 90°C. were necessary to obtain a paste of yeast cells free from hydrocarbons.

(6) (5) was repeated carrying out the washings with water at 60°C. containing 0.5 part per1,000 of Cemulsol. Two washings at 60°C. were sufficient to obtain a paste of the cells free from hydrocarbons. Finally, the yeast was washed with pure water at 60°C. to eliminate the surface active agent remaining in the yeast.

Example 2: The emulsion from which it was desired to extract the yeasts was constituted by cells of *Candida lipolytica* and paraffinic gas oil finally dispersed in small particles in a continuous aqueous phase of a mineral medium identical to that of Example 1 and having a pH of 3.5. Experiments were carried out on samples taken from the original emulsion.

(1) Intermittent centrifuging at 60°C. of the original emulsion gave a separation as imperfect as that of (1) of Example 1.

(2) The pH of the original emulsion was adjusted to 8 and centrifuged at 60°C. The separation was a little better but still quite insufficient.

(3) There was added to the original emulsion 0.25 part per 1,000, with reference to the aqueous phase, of the cationic surface active agent stearyl-trimethyl ammonium chloride and the pH adjusted to 8. The centrifuging was carried out at 60°C. and gave the same three phases which were clearly separated as in Example 1, (4).

(4) The pasty phase of yeast cells impregnated with hydrocarbons from (3) was washed with water at 60°C. and then centrifuged. Five successive washings and centrifugings were necessary to obtain a paste of yeast cells free from hydrocarbons.

(5) (4) was repeated carrying out the washing with water containing

0.25 part per 1,000 of the same surface active agent as in (3). Two washings and centrifugings were sufficient to obtain yeasts free from hydrocarbons. A final wash with pure water at 60°C. was carried out to eliminate surface active agent remaining in the yeast.

Coagulation by Low pH and Heat

J.W. Frankenfeld, R.P. Cahn and P. Zayhekian; U.S. Patent 3,427,223; February 11, 1969; assigned to Esso Research and Engineering Co. have reported on coagulating microbial cells by adding to an aqueous effluent of a culture media either a cell nutrient containing acid or mixtures of the nutrient containing acid and a cell nutrient containing salt, sufficient to lower the pH in the range of 2.0 to 6.0, thereafter heating for sufficient time to coagulate the cells and thereafter separating the coagulated cells from the effluent.

One serious drawback to the wide acceptance of protein biosynthesis techniques, especially when hydrocarbons are employed as a predominate feed media, resides in the fact that the product microorganism cells, especially in the case of certain bacteria cells, are often of very small size, e.g., from 0.5 to 5.0 microns and even smaller.

With such small sized microorganism cells it is difficult and often very expensive to achieve recovery of the product microorganism cells. One factor leading to increased expense in recovering product microorganism cells is the difficulty in separating the cells from the aqueous product stream containing them where the cells are present in low concentrations, e.g., 1 to 5 wt. percent.

Another salient factor which increases recovery expense is the required removal of comparatively large amounts of water attracted to and maintained by the cells, including interstitial water present in the cell walls of the microorganisms. Still another cost increasing factor resides in the apparent necessity to pasteurize the cells, viz, render them essentially nonviable. These and other related factors serve to increase the overall cost of recovering the grown microorganisms.

Separation and dewatering, e.g., drying, procedures can amount to as much as 25% or higher of the total cost of producing the microorganisms in a form suitable for use in high protein foods or food supplements. Hence, any substantial improvement in effecting dewatering of microorganism cells especially those having very small cell diameters, prior to separation and drying will result in significant economic savings and thus reduce the overall cost of producing high protein foods and food supplements.

In this process, the rate at which the microbial cells can be separated from the remainder of the product stream can be accelerated by as much as 20-fold or even higher, thus attaining substantial economies in time and equipment necessary for conducting the separation procedure on a given volume of product stream.

The substantial reduction in the time required for separation and drying of the cells is usually applicable regardless of what separation technique is employed. For example, substantial savings in the time and equipment required for effecting separation and dewatering of microbial cells can be achieved using any of the conventional cell separation techniques, e.g., filtration, centrifugation, settling, etc.

Another advantage is that the process allows the same separator and drying apparatus to treat a much larger volume of product containing aqueous effluent stream within a given period of time than could be handled previously.

For example, when the speed of filtration is increased by a factor of four-fold, the filtration apparatus can handle four times the volume of product containing aqueous effluent than could be filtered in the same given time period. Similar economies of time and efficiency in utilization of separation apparatus are achieved when the separation procedure is conducted by centrifugation.

The effluent stream containing microbial cells is withdrawn from the biosynthesis bath. The pH of the product stream effluent is then lowered by addition of a nutrient containing acid or acid plus salt. The term "nutrient containing" is employed in a dual meaning in that the material in question must not only contain a cell nutrient, but also must be free from a component(s) which are detrimental to cell growth and reproduction.

Hence, acids or salts having a poisonous effect on the microorganisms being harvested must be avoided. Then the acidified effluent is heated prior to passing it through a separator which separates the microbial cells from the remaining portion of the acidified effluent.

The cells are withdrawn from the separator and subjected to drying to produce the dried product microbial cells. The remaining portion of material coming from the separator, viz, the filtrate or supernatant containing the previously added nutrient acid is then recycled back into the biosynthesis bath.

Prior to or after feeding the nutrient acid containing filtrate or supernatant back to the biosynthesis bath, its pH is raised to a level closely approximating that maintained for cell growth in the biosynthesis bath.

After addition of the nutrient containing acid, the acidified effluent is preferably heated. The heating is conducted at a temperature above the growth temperature maintained in the biosynthesis bath yet below the temperature at which cell degradation occurs (protein degradation). Temperatures can be used which cause protein denaturation as long as no degradation of food value occurs. The heating is continued for a sufficient period of time to attain substantial coagulation of cells.

The use of heating after acidification but before cell separation is preferred as it contributes substantially to cost reductions and efficiency. One advantage noted is a substantial improvement in acceleration of the rate at which cell separation can be conducted; and this improvement is specially pronounced in an area of difficult cell separations, viz, where the individual cell size is about one to two microns or below.

Another important advantage is the attainment of more cell solid, viz, less residual water, in the separated cells. This reduces the cost of drying the separated cells, and drying is usually regarded as the most expensive step in cell recovery.

Still another advantage resides in the partial sterilization of the microorganism cells. In this context, cells can be rendered partially nonviable with the use of substantially lower pasteurization temperatures in the acidified aqueous effluent stage.

In order to avoid degradation of cells during the coagulation step, the temperatures to which the acidified effluent is heated are usually controlled to range from a temperature about 10°C. above the growth temperature maintained in the biosynthesis bath up to a temperature below that at which cell degradation occurs, e.g., a temperature of approximately 100°C. or even below. The exact upper and lower temperatures will vary depending upon the specific microorganism being grown. Usually, the acidified effluent is heated to a temperature ranging from 45° to 90°C. for a time period ranging from 1 to 30 minutes, and more preferably from 1 to 20 minutes. For example, heating times of 1 to 15 minutes at temperatures ranging from 50° to 85°C. usually yield excellent coagulations for most bacterial cells, including *Micrococcus cerificans.*

In the case of most bacteria cells, it is usually advisable to acidify the product effluent stream issuing from the biosynthesis bath to a pH ranging from 2.5 to 5.5. For example, the pH of the bacteria cell containing aqueous effluent is lowered to a pH ranging from 3.0 to 4.5 when *Micrococcus cerificans* bacteria cells are being coagulated.

Example 1: Micrococcus cerifans is continuously grown in an aqueous biosynthesis bath containing 1.0 weight percent n-hexadecane as the hydrocarbon feed media.

A 7.5 liter biosynthesis reactor vessel is charged with a 4 l. aqueous slurry containing 1.0 weight percent *Micrococcus cerifans* bacteria as inoculant. Sufficient air is bubbled through the inoculant slurry to satisfy the bacterial oxygen demand prior to introducing the hydrocarbon and inorganic salts growth feeds (the latter containing phosphoric acid and ammonium hydroxide). A typical biosynthesis bath composition at a given typical stage of a continuous biosynthesis campaign is as follows:

Component	Grams per Liter
n-Hexadecane	10
H_3PO_4	5
KCl	1
$CaCl_2$	0.5
$MgSO_4 \cdot 7H_2O$	0.2
$MnSO_4 \cdot 4H_2O$	0.2
$FeSO_4 \cdot 7H_2O$	0.2
NaCl	0.2

Sufficient NH_4OH is added to maintain the pH at 7.0. The biosynthesis bath growth temperature is maintained at 35°C. (95°F.), plus or minus 2°C., and the pH of the biosynthesis bath is maintained essentially neutral, viz, at a pH of 7.0, plus or minus 0.1, throughout biosynthesis.

The conversion of hydrocarbon feed to cells is maintained at 90% and higher. After a residence time of approximately 2 hours, a product stream aqueous slurry is continuously withdrawn from the biosynthesis bath. Typical 100 milliliter samples of this product stream aqueous slurry, containing approximately 1% by weight of bacteria cells, along with unconverted hydrocarbon, inorganic salts, nutrients, etc., are withdrawn at varying stages of a biosynthesis campaign. These samples are divided into four groups for further processing.

One group of samples, Group 1, is filtered directly without coagulation pretreatment. Another group of samples, Group 2, is acidified to a pH of 3.5 by addition of phosphoric acid prior to filtering. Another group of samples, Group 3, is acidified to a pH of 3.5 by the addition of phosphoric acid and then heated at a temperature of 85°C. (185°F.) for 15 minutes prior to filtering. The remaining group of samples, Group 4, was heated at 85°C. for 15 minutes but not acidified prior to heating and filtering.

Following this all samples are gravity filtered through the same filter, viz, 11 cm. fluted Whatman #1 filter paper. The average rates of filtration for the samples of each group are tabulated in the table below. In the table, both the pH and the temperature data are taken at the time of filtration.

Group	Acid Treatment	Heat Treatment	pH	Temperature, (°C.)	Filtration Rate ml./min.
1	No	No	7.0	25	0.53
2	Yes	No	3.5	25	2.22
3	Yes	Yes	3.5	85	10.00
4	No	Yes	7.0	85	0.16

As will be noted from the above data, the process enables the attainment of filtration rates approximately 19 times faster (Group 3) than when no coagulation treatment is conducted.

Following filtration, the phosphoric acid containing filtrate is recycled back to the biosynthesis bath to supply makeup phosphorus. The pH of the biosynthesis bath is controlled to maintain a pH of approximately 7.0 continuously by addition of ammonium hydroxide when needed.

Example 2: Micrococcus cerificans cells are grown as in Example 1 to yield an aqueous slurry product stream containing approximately 1% by weight cells. Typical 100 ml. samples of this product stream are withdrawn and divided into five groups for various coagulation treatments. Phosphoric acid was used to bring the pH of all samples to 3.5, and the samples of each group were then heated for 15 minutes at various temperatures. The samples were then gravity filtered as in Example 1 to yield the average filtration rate as shown below.

Group	Temperature, (°C.)	Filtration Rate (ml./min.)
1	50	2.54
2	60	3.9
3	70	4.7
4	80	5.1
5	85	10.0

Example 3: Micrococcus cerificans cells are grown as in Example 1 to yield an aqueous slurry product stream containing approximately 1% by weight cells. Typical 100 ml. samples of this product stream are withdrawn and divided into 5 groups for filtration.

None of the samples are heated, but instead all samples are filtered at a temperature of 25°C. (77°F.) using gravity filtration as in Example 1. This temperature is approximately 10°C. below the growth temperature maintained during biosynthesis. The pH of the samples in each group was adjusted using phosphoric acid to yield the pH values shown below. Despite wide variance in pH the filtration rates attained were slower than secured when adequate heating was used following filtration.

Group	pH at Time of Filtration	Filtration Rate (ml./min.)
1	2.0	0.42
2	3.0	1.7
3	3.5	2.2
4	4.0	1.25
5	5.0	0.63

Phase Separation by Alcohol and Hydrocarbon

In the procedure of *J.B. Bonavita; U.S. Patent 3,655,511; April 11, 1972; assigned to The British Petroleum Company Limited, England,* a product mixture consisting of a microorganism, unconsumed hydrocarbons and an aqueous phase is treated with an alcohol and a solvent hydrocarbon used in either order or together. Each part by volume of product mixture is treated with 0.5 to 2 parts by volume of alcohol and with 0.5 to 2 parts by volume of solvent hydrocarbon. The treated product mixture is thereafter subjected to phase separation with recovery of a phase comprising the solvent hydrocarbon and unconsumed hydrocarbons and with recovery of a phase mixture comprising alcohol, water and the microorganism.

The phase mixture recovered is subjected to phase separation to recover a solid phase product consisting of the microorganism. This is achieved by filtering the phase mixture comprising the alcohol, water and microorganism. A filter cake so obtained is washed with water or an aqueous medium to remove traces of the alcohol. If desired, other phase separation methods can be employed, for example centrifuging.

Liquid fractions obtained by the process are treated, for example by distillation, singly or after blending of two or more fractions, to recover the solvent hydrocarbon and, separately or in admixture with alcohol.

The relative proportions of alcohol and solvent hydrocarbon which are employed in the treatment of the product mixture lie in the range 1 to 4 parts by volume of solvent hydrocarbon per part by volume of alcohol.

The phase mixture comprising alcohol, water and the microorganism which is subjected to phase separation contains less than 20% by volume of water. It is believed that this limitation on water content has a significant effect on the ease of subsequent separation of the microorganism from the liquid phase.

Suitably the solvent hydrocarbon is selected from pentanes, hexanes and mixtures of hydrocarbons, for example, petroleum ethers. However, the preferred solvent hydrocarbon is normal hexane.

Suitably the alcohol is selected from ethanol, propanol, isopropanol and the butanols. The preferred alcohol is isopropanol. Usually the straight chain hydrocarbons employed in the feedstock, for cultivation of microorganisms, will be paraffins; however, the straight chain hydrocarbons may be present as olefins; also there may be used a mixture containing straight chain paraffins and olefins.

Suitable feedstocks to the process include kerosene, gas oils and lubricating oils; these feedstocks may be unrefined or may have undergone some refinery treatment, but must contain a proportion of straight chain hydrocarbons in order to fulfil the purpose of this process. Suitably the petroleum fraction will contain 3 to 45% by weight of straight chain hydrocarbons. Preferably the petroleum fractions have a mean molecular weight corresponding to at least 10 carbon atoms per molecule.

The process is of particular value for the treatment of petroleum gas oil fractions which contain straight chain hydrocarbons in the form of waxes, since by the process a gas oil of improved pour point is obtained while the waxes are converted to a valuable product.

Example 1: An aqueous mineral medium was made up of the following ingredients.

	Grams/Liter
$(NH_4)_2HPO_4$	2
KCl	1.15
$MgSO_4 \cdot 7H_2O$	0.32
$MnSO_4 \cdot 7H_2O$	0.022
$ZnSO_4 \cdot 7H_2O$	0.085
$FeSO_4 \cdot 7H_2O$	0.034
Yeast Extract	0.015

2.6 liters of this mixture was introduced into a 3 liter stirred fermenter. To the fermenter was added as an inoculum, 1.5 grams of the yeast *Candida tropicalis* in 200 ml. of the above aqueous nutrient medium.

The carbon substrate was a gas oil which had a specific gravity (at 60°F.) of 0.865, a cloud point of +18°C., a pour point of +16°C., contained 0.90% by weight sulfur and 15% by weight normal paraffins.

Continuous cultivation was carried out with addition of the carbon substrate and aqueous nutrient medium under the following conditions.

Carbon substrate feed rate	0.1 vol./vol./hour
Dilution rate	0.2 vol./vol./hour
Agitation by stirring - rate	2,500 rpm
Aeration	30 vol./vol./hour
pH*	5.5
Temperature	30°C.

*pH control was by addition of 5N aqueous ammonia.

The microorganism product was recovered as follows. The broth obtained from the fermenter was continuously fed to a decanter with removal of a lower aqueous phase and recovery of an upper phase containing the yeast, hydrocarbons and some aqueous material. The upper phase constituted 25% by volume of the broth.

To 100 ml. of the upper phase was added with stirring 100 ml. of isopropanol followed by 100 ml. of normal hexane. Two liquid phases separated, the upper phase having a volume of 145 ml. and containing the bulk of the normal hexane, with residual hydrocarbon from the carbon substrate and lipids extracted from the yeast.

The upper phase was separated by decantation. The lower phase containing water, the bulk of the isopropanol and yeast in suspension was filtered on a Tergal lacquered filter at reduced pressure with recovery of a solid microorganism product substantially free of hydrocarbons and solvent. Four grams of dry microorganism were obtained.

Example 2: The process described in Example 1 was repeated except that the amounts of isopropanol and normal hexane which were employed were respectively 150 ml. and 50 ml.

Four grams of dry microorganism were obtained. In comparison with Example 1, the decantation rate was found to have been assisted by the use of the greater quantity of isopropanol but, as a result of the use of the lower quantity of hexane, the recovered biomass was found to contain trace amounts of hydrocarbon.

Phase Separation by a Dialkyl Ketone and Acetone or Alcohol

Yeast fermentation worts containing hydrocarbon, water, and inorganic salts, are resolved by contacting the worts with a mixed solvent system containing a dialkyl ketone (a) having at least four carbon atoms, e.g., methyl ethyl ketone, and another solvent (b) selected from acetone, ethanol, n-propanol, isopropanol and tert-butanol.

As a result of this technique developed by *C. Gatellier and G. Glikmans; U.S. Patent 3,625,832; December 7, 1971; assigned to Institut Francais du Petrole, des Carburants Et Lubrifiant, France* three phases appear: a heavy aqueous phase containing the microorganisms, for example yeasts, an intermediate phase containing water and the solvent of group (b), and a lighter phase which is rich in solvent of group (a) and residual hydrocarbons.

The effluent from the fermentation vessel is preferably introduced at first into a quiet zone in order to obtain a first separation in two phases, one consisting of the so-called emulsion containing the residual hydrocarbons, the yeasts and a portion of the aqueous inorganic medium, the other mainly consisting of the aqueous inorganic medium.

If a large amount of hydrocarbons remains unconverted at the end of the culture (for example when as substrate is used a gas oil containing a minor amount of straight chain paraffins), three phases may appear: a hydrocarbon phase, an aqueous phase and the emulsion phase containing water and a large amount of residual hydrocarbons.

Although the aqueous phase may be recycled to the fermentation vessel, the emulsion is collected in order to be subjected to treatment.

It must be pointed out that this separation in three phases is obtained only with the above mixture of solvents. Thus when there is used for example a mixture containing no solvent of the group (a), for instance a mixture of acetone and isopropanol, the formation of three separate phases does not occur.

In this process yeasts may be separated from the fermentation wort without use of centrifugations whereby the cost of the process is substantially reduced. At the same time the microorganisms are washed by the solvents so as to make them free or substantially free from hydrocarbons and organic by-products resulting from the fermentation, for example fatty acids, fatty alcohols or esters.

It is thus easy to recover, for example by mere decantation and withdrawal, the heavy phase of yeasts from the three phases obtained by this process.

The two other phases may also be easily separated, and, after removal, for example by distillation, of the residual hydrocarbons of the lightest phase and water of the intermediary phase, the solvents of groups (a) and (b) may be easily reused in a separation and purification step.

This treatment may be carried out in several manners. For example, the emulsion may be contacted with the solvent of group (a) and then with that of group (b), or otherwise the emulsion may be contacted with the preformed mixture of (a)+(b). If desired, the paste of yeasts obtained by decantation may be thereafter contacted with a mixture of solvents which is identical to that used for separating the yeasts from the fermentation medium.

The total amount of solvents (a) and (b) to be contacted with the emulsion from the yeasts cultivation zone, and more particularly the ratio of the solvents (a) and (b) are dependent on the amounts of aqueous and hydrocarbon phases of this emulsion.

It is preferred to use a small excess of solvent of type (a) with respect to the solvent of type (b); for example the ratio by weight of the solvent of type (a) to the solvent of type (b) will be between 1 and 8 and more particularly between 1.5 and 5, although these values are not considered as limitative.

As a rule, there will be used from 1 to 10 and preferably from 2 to 5 volumes of mixture of the solvents (a) and (b) per volume of emulsion containing the microorganisms.

Example 1: Yeasts of the *Candida lipolytica* type are continuously grown in a fermentation vessel, in the presence of a carbon source consisting of a gas oil containing 11% of linear paraffinic hydrocarbons and in the additional presence of an aqueous nutritive phase, the composition of which is given hereafter.

Air is injected in a divided form from the bottom of the vessel. The pH is maintained at about 4 and the temperature at about 27°C. The aqueous nutritive medium has the following composition.

	Grams
Diammonium phosphate	2.2
Potassium chloride	1.1
Magnesium sulfate	0.9
Tap water (containing oligo-elements)	320
Yeast extract (growth factor)	0.03
Distilled water	1,000

The liquid effluent from the fermentation vessel is passed into a decantation vessel in which a first separation takes place between an aqueous phase, half of which is recycled to the fermentation vessel, a phase of unconverted gas oil and an emulsion or paste of yeasts consisting essentially of water (with inorganic salts), in a proportion of about 90% by weight, and of 5% of yeasts and 5% of residual hydrocarbons.

One volume of this paste is contacted with 2 volumes of a liquid mixture consisting of 4 parts by volume of methyl ethyl ketone and one part by volume of acetone (the ratio by weight is also about 4/1).

Three phases separate: a lower aqueous phase of yeasts, an intermediary phase essentially comprising acetone and water and a lighter phase mainly comprising methyl ethyl ketone and residual hydrocarbons.

The aqueous phase is withdrawn; the yeasts contained therein are free of residual hydrocarbons and of by-products of the fatty acid type. They may be dried and stored for a long time without turning rancid. The two other phases are distilled, and the thus recovered acetone and methyl ethyl ketone may be used anew.

Examples 2 and 3: Example 1 is repeated with the following changes. There is used a mixture of methyl isobutyl ketone and ethanol (Example 2) and a mixture of methyl ethyl ketone and propanol (Example 3).

All other conditions remaining unchanged, particularly the amounts of solvents, there is obtained a separation of the emulsion in three phases from which yeasts substantially free from hydrocarbons and by-products may be obtained.

Example 4: Example 1 is repeated with the following change. The hydrocarbon feed charge for the culture step has been obtained by dewaxing a lube oil. This charge contains 90% of straight chain paraffins. The aqueous nutritive medium remains unchanged.

By decantation of the effluent from the fermentation vessel, there is obtained an emulsion consisting of 90% by weight of water, 9.5% of yeasts and 0.5% of residual hydrocarbons.

One volume of this emulsion is contacted with 5 volumes of a mixture consisting of 1.5 parts by weight of methyl ethyl ketone per part by weight of acetone. As in Example 1, three phases appear. The yeasts are removed from the heavy phase; they are substantially free from residual hydrocarbons and by-products.

Phase Separation in a Gravitational Field

A process for treating oil containing fermentation discharges has been described by *M. Rudel, A. Gabert, U. Behrens, D. Pohland,*

M. Ringpfeil and K. Sattler; U.S. Patent 3,578,566; May 11, 1971; assigned to Deutsche Akademie der Wissenschaften zu Berlin, Germany.

An oil containing fermentation discharge such as obtained in the aerobic fermentation of petroleum fractions and petroleum-like paraffinic products is subjected in the form of a dispersion of the oil and microorganisms in the nutrient medium to a gravitational field of a sufficient magnitude and for a time sufficient to effect a phase inversion.

The fermentation medium and microorganisms in the inverted phase are dispersed in the oil so that the latter forms the external phase of the dispersion. At least part of the oil is separated from the microorganisms. Preferably the microorganism is yeast and the gravitational force is applied by means of a centrifugal field having an acceleration in excess of $10^3 g$; g being the gravity acceleration.

It will be understood that the magnitude of the centrifugal acceleration depends also on the period of time for which the treatment is carried out. It is possible to obtain the desired result with smaller values of g if the process is carried on for a longer period of time.

The required time of treatment and the magnitude of the centrifugal acceleration can further be varied by varying the temperature of the fermentation discharge. An increase of temperature will have a positive effect on the speed at which the phase inversion is reached. An upper limit for the process conditions, is set in the practical application by the point where damage to the microorganism occurs.

The suspension or emulsion phase formed with oils as external phase was rather unstable against additional mechanical action. Thus substantial amounts of the external oil phase could be separated until finally an extreme limit value of emulsion was obtained. This borderline emulsion comprised between about 5 and 25 parts of yeast, 75 and 30 parts of water and 20 to 45 parts of oil. The mechanical action referred to may consist in stirring, filtration or similar operations.

The emulsion could be broken for good also prior to removing the oil by withdrawing water therefrom. This could be done with the suspension or emulsion first obtained wherein the oil formed the external phase or it could be done in addition to prior mechanical action as just described. The removal of water can be accomplished by use of hydrophilic solvents or the water can be eliminated by evaporation preferably in vacuum. To break the emulsion it is also possible to add highly soluble salts.

The duration of the treatment, the temperature and the amount of acceleration will of course depend on the composition of the discharge. They may all be varied within a broad range. The time may for instance be between 1 second to 10 minutes, the temperature may vary between +4° to about 90°C. and the acceleration preferably should be above $10^3 g$ as above stated with the upper limit being the danger of injury to the microorganisms.

A preferred relationship would be a treatment for about 10 minutes with a centrifugal field of 5.5 times $10^3 g$ at room temperature. The following example will illustrate the process without limiting its scope:

68.41 g. of the light phase of a discharge obtained from the aerobic fermentation of diesel oil with a yeast of the genus *Candida lipolytica* was subjected for 10 minutes at room temperature to a centrifugal field of the magnitude 5.5 x $10^3 g$, g being the gravitational acceleration. Through this treatment separation of 26.3% water was obtained which then was eliminated.

The formed product was subjected to stirring in a beaker resulting in the separation of oil. In this manner 88.5 weight percent of the oil which was originally present in the lighter phase could be separated.

The pasty residue amounting to about 20 g., was subjected to vacuum distillation in a rotary evaporator at 100 torr (287.5 pound/foot square). There was formed a distillate amounting to 13 g. The yeast was obtained as a dry oil-containing product which could then be freed of the oil in a conventional manner.

REMOVAL OF RESIDUAL HYDROCARBON

Application of a Water-Soluble Surfactant

B.M. Laine; U.S. Patent 3,536,585; October 27, 1970; assigned to The British Petroleum Company Limited, England has grown microorganisms on a hydrocarbon feedstock and then separated from the cultivated mixture, through the use of a surfactant, a fraction containing, as the major proportion unconsumed hydrocarbons together with a minor proportion of water and a minor proportion of surfactant.

Upon heating the fraction, the surfactant is rendered partially or substantially water-insoluble, a hydrocarbon phase containing surfactant is separated from an aqueous phase and the hydrocarbon phase is hydrogenated under conditions such that the surfactant is rendered less effective or ineffective as an emulsifying agent.

When a water-soluble surfactant is employed in the hydrocarbon recovery stage the separation of the bulk of the aqueous phase from the hydrocarbon phase is much assisted but then remains a problem of breaking an emulsion of hydrocarbon and water which is also recovered.

This emulsion can be effectively broken by heating; however, in consequence the surfactant is transferred to the hydrocarbon phase. By the overall process, which includes a hydrogenation stage, the modified surfactant is reduced in effectiveness or is rendered ineffective.

After growth it will usually be possible to separate the microorganism, contaminated with some unmetabolized feedstock and aqueous nutrient medium, from the bulk of the unmetabolized feedstock fraction.

Preferably the separation is achieved by means of a decantation; additionally or alternatively centrifuging may be used. The fraction containing the microorganism is subjected to treatment with an aqueous treating medium comprising water, a metal salt and a surface active agent consisting of or containing a nonionic detergent.

Example: 40 liters of an aqueous mineral nutrient medium having the composition given below were introduced in a stainless steel fermenter having an effective capacity of 60 liters.

In order to keep the temperature in the fermenter constant at 30°C., water was circulated in an annulus constituted by the space between two concentrical cylinders, the smaller one being the fermenter itself.

20 liters of a 24 hours inoculum of *Candida tropicalis* on mixed C_{10} to C_{20} normal hydrocarbons were added to an aqueous medium

(composition shown below) such that the cellular density was about 1 gram of dry matter per liter.

	Grams
Diammonium hydrogen phosphate	2.
Potassium chloride	1.15
Magnesium sulfate · $7H_2O$	0.65
Zinc sulfate	0.17
Manganese sulfate · H_2O	0.045
Ferrous sulfate · $7H_2O$	0.068
Yeast extract	0.025
Tap water	200
Distilled water q.s. 1,000 ml.	

1,030 liters of heavy gas oil, that is 15 grams/liter, were then added this quantity being sufficient to take the cellular density to 2 grams/liter.

The temperature of the culture was kept at 30°±1°C. pH at 4 and aeration and agitation giving 3 millimols of O_2 per liter of medium per minute. Ammonia solution was admitted by an automatic pH controller.

When the flow of ammonia reached 20 ml. the addition of gas oil was begun assuming a yield on gas oil (dry yeast produced/gas oil feed) of 10% by weight and a cell division time of 3 hours. This addition was carried out every hour until 200 grams/liter, i.e., 13.8 liters had been added.

Starting with a cellular density of 2 grams/liter, at 25 hours (at the end of the exponetial growth phase) the cellular density was 15 grams/liter. The fermenter was then run continuously at a dilution rate of 0.2 vol./vol./hour, cellular density remaining constant at 15 grams/liter throughout the run. The broth collected at the outlet of the fermenter was continuously withdrawn from the fermenter and subjected to decantation, 65% of spent medium being withdrawn and replaced by 65% of tap water.

To the upper phase was added 0.5 gram/liter of the nonionic detergent (NI29) and, after centrifuging, there were separately recovered:

Spent mineral medium	839 g./l.
Nonmetabolized gas oil	112 g./l.
Paste of microorganism	49 g./l.

The nonmetabolized gas oil had a specific gravity (60°F.) of 0.8803; an n_D^{20} of 1.490; percent by weight of sulfur, 1.84; a cloud point of +1°C; a pour point of −6°C; a trace of water; and a trace of detergent.

The trace amount of detergent was found to have led to undesirable emulsifying characteristics in the gas oil. This gas oil was tested for emulsion forming characteristics by the test method known as ASTM B.1400.01 56T. In this test 40 cc of water is used, 8 cc of kerosene (as diluent) and 32 cc of the gas oil undergoing test. Results were as shown in the following table.

The recorded gas oil containing trace amounts of water and detergent was heated to 90°C., whereby the detergent was dissolved in the oil phase, and centrifuged at 90°C. in a Sharples tubular centrifuge at 14,000 g. to remove water.

The residual gas oil so obtained was hydrogenated in the presence of a cobalt molybdate catalyst containing: 12.6% by weight MoO_3 and 2.45% by weight of CoO.

200 cc of the catalyst was used in the form of 2.5 x 2.5 mm. cylindrical pellets and hydrogenation was carried out at 395°C. and 40 kg. pressure.

The gas oil throughput rate in vol./vol./hr. was 3 in Run 1, 5 in Run 2 and 10 in Run 3. The hydrogen throughput rate in l./hr. was 50 in Run 1, 85 in Run 2 and 180 in Run 3. The treated gas oil was tested for emulsion forming characteristics by the test method ASTM B.1400-01 56T. In this test 40 cc of water is used, 8 cc of kerosene (as diluent) and 32 cc of the gas oil undergoing test. Results are shown in the table.

Emulsification Test	Feedstock*	Recovered Gas Oil*	Hydrogenated Gas Oil*		
			Run 1	Run 2	Run 3
Separation time	30 sec.	>60	13 sec.	70 sec.	74 sec.
Oil (cc)	41	1	39	41	40
Water (cc)	38	27	40	38	38
Emulsion (cc)	1	52	2	2	2

*Kerosene diluted.

Use of a Two-Component Combination of Surface Active Agents

In the procedure of *U. Behrens, M. Ringpfeil, A. Gabert, D. Pohland, K. Sattler, M. Rudel, V. Munk and J. Jiricka; U.S. Patent 3,616,248; October 26, 1971; assigned to Deutsche Akademie der Wissenschaften zu Berlin, Germany* the yeasts in yeast-oil-water emulsions formed in the conversion of hydrocarbons to yeast are separated from the emulsion by adding one of the following two-component combinations of surface-active agents to the emulsion prior to the separation step:

(1) Polyphenol-ethylene oxide adduct (containing about 16 mols of ethylene oxide per mol of polyphenol) and olein-monoethanolamide.

(2) Polyphenol-ethylene oxide adduct and distillation-residue of the fatty acid distillation C_{12} to C_{18}.

(3) Polyphenol-ethylene oxide adduct and hydroxyethylated sulfated fatty alcohol having a chain length between C_{12} and C_{18}.

(4) Alkylomonosulfonate (mean chain length of about C_{15}) and laurylamidoethyl pyridinium chloride.

(5) Alkylmonosulfonate (mean chain length of about C_{15}) and the methosulfate of a quarternary fatty amine having a mean chain length of between C_{15} and C_{17}.

The surface active agents are used in an amount up to 0.15%, preferably below 0.1%, relative to the total emulsion. The following technique may be observed for the operation.

During the conversion of the hydrocarbon to yeast, a separation occurs under influence of gravity or in a centrifugal operation into an aqueous lower phase and a creamy emulsion phase. Virtually all of the yeast components of the fermentation solution are found in the emulsion phase together with a corresponding amount of water and of unreacted hydrocarbon.

Depending on the yeast and oil content, the emulsion has a more or less pronounced creamy character. To this emulsion there is added at first only one component of the two-component combination. This component is thoroughly mixed with the emulsion. Thereafter the second component is added and mixed into the emulsion. The mixture, which has a substantially reduced viscosity compared with the initial emulsion, is then passed to the separator. The temperature of the mixture subject to the separation should not be in excess of 40°C. if the combination of surface-active agents involves a cationic and an anionic substance. If nonionic surface-active agents are used, it is preferable to increase the temperature, for instance to between 40° and 60°C.

The use of the combination of surface-active agents permits a more or less quantitative separation of the emulsion into yeast-water and oil.

Example 1: 100 g. of a yeast-oil-water emulsion are reacted and thoroughly mixed with 0.025% of a polyphenol-ethylene oxide adduct in the form of a 1% aqueous solution 0.05% of olein-monoethanolamide are thereafter added and mixed into the emulsion. The mixture is heated for half an hour to a temperature of 40°C. and then subjected to centrifuging for 5 minutes at 3,000 rpm. Separation into yeast-water-oil is thus effected.

Example 2: 100 g. of a yeast-oil-water emulsion are mixed with 0.01% of polyphenol-ethylene oxide adduct as 1% aqueous solution. There is then added 0.1% of the distillation residue of a fatty acid distillation of acids with a mean chain length C_{12} to C_{18}. The mixture is warmed for half an hour to 60°C. and subsequently centrifuged for 5 minutes at 2,000 rpm. Separation into yeast-water-oil is effected.

Example 3: 100 g. of a yeast-oil-water emulsion are mixed with 0.01% of polyphenol-ethylene oxide adduct as 1% aqueous solution and are then reacted with 0.025% of a hydroxyethylated sulfated fatty alcohol of a mean chain length C_{12} to C_{18} in the form of a 1% aqueous solution. The emulsion is thoroughly mixed with the additives and warmed up for half an hour to 40°C., followed by centrifuging. A separation into yeast-water-oil is effected.

Refermentation Techniques and Solvent Extraction

J.A. Filosa; U.S. Patent 3,520,777; July 14, 1970; assigned to The British Petroleum Company Limited, England has described the cultivation of a microorganism in the presence of a hydrocarbon feedstock consisting wholly or in part of straight chain hydrocarbons, in the presence of an aqueous nutrient medium and in the presence of a gas containing free oxygen. The microorganism is maintained in a purification stage with an aqueous nutrient medium and a gas containing free oxygen whereby the hydrocarbon contaminating the microorganism is reduced in quantity. The product containing the microorganism is subjected to a separation stage recovering a fraction containing the microorganism. The fraction containing the microorganism is subjected to solvent extraction and the fraction containing the microorganism obtained from the solvent extraction stage is dried.

According to one method of treating the product the major part of the continuous aqueous phase is first separated; preferably this is carried out by centrifuging, or decanting. The separated aqueous phase will usually contain a greater concentration of nonnutritive ions than can be tolerated in the recycle stream and when this is so, only a proportion of the recovered aqueous phase can be recycled.

Thus it will usually be possible to separate about 96% by weight of the aqueous phase which is present in the product, of which on the same percentage basis, about 20% by weight will be discarded. The recycle stream is supplied with makeup quantities of the necessary nutrients and is returned to the fermenter; if desired the makeup materials may be fed to the fermenter as a separate stream.

By centrifuging the product from the fermenter (preferably after decanting as described) three fractions are recovered. These are in order of increasing density:

(1) An oil phase containing microorganism cells;
(2) an aqueous phase containing traces of oil and microorganism, and
(3) a microorganism "cream" consisting of microorganism, having a quantity of oil fixed on to the cells, together with aqueous phase.

Fraction (3) will then be mixed with an aqueous nutrient medium, which may be the same as or different from the aqueous nutrient medium employed in the growth stage, and is maintained in mixture with a gas containing free oxygen whereby the hydrocarbon contaminating the microorganism is reduced in quantity or is eliminated.

Again, preferably, the major part of the continuous aqueous phase is separated with recycle of a proportion of the recovered aqueous phase, as hereinbefore described; recycle may be to either the growth stage or purification stage as desired. By centrifuging the product from the fermenter, preferably after separating the major part of the aqueous phase by decanting, there are obtained:

(1) An aqueous phase containing traces of microorganism;
(2) a microorganism cream.

This microorganism cream which contains aqueous material may be treated by continuous solvent extraction or by successive washings with solvent followed by phase separation. Preferably the cream contains 1:1 to 3:1 by weight of water to microorganism (dry weight). If desired a paste may be recovered by centrifuging and diluted with water to form the cream.

Suitably the extraction is carried out in a stationary vessel equipped with paddle stirrers, preferably rotating at less than 10 rpm or in a vessel which rotates on a horizontal axis. When operating a continuous solvent extraction, the extract is withdrawn continuously and distilled, continuously or batchwise, at atmospheric or reduced pressure, and solvent continuously fed back to the extractor. Under these conditions the yeast may be introduced and withdrawn continuously or batchwise.

Preferably the solvent extraction is effected while feeding solvent to the extractor at a periodically varying rate to create pulsations in the flow of the liquid stream.

The pulses of the liquid passing through the solid material bring about oscillations and limited displacements of each grain of solid material in relation to its neighbors, and this is equivalent to a mechanical agitation of the whole. For this reason the whole of the products to be extracted is much more rapid and complete.

Suitably there is arranged in the feed of the liquid stream a device which imparts to it pulses whose amplitude and frequency are regulated experimentally at the most favorable value for each particular case. These pulses are produced by any suitable processes already known, and preferably an alternating pump is used whose valves have been removed.

Suitable solvents for use in the process have been described hereinbefore. If desired a first extraction stage can be operated using a polar solvent, for example an alcoholic solvent, for example ethanol or isopropanol and then the partially purified microorganism can be further treated in a second extraction stage using a hydrocarbon solvent, for example normal hexane or a light platformate fraction or benzene. Preferably in the second stage there is used as solvent a mixture of hydrocarbon in major amount with a polar solvent in minor amount. Preferably there is used the azeotropic mixture of hexane with isopropanol or ethanol. If desired both extraction stages can be operated in continuous manner.

When using a solvent consisting of a mixture of hydrocarbon and a polar solvent, it is believed that the function or one function of the polar solvent is to weaken the bonding of the material to be extracted (even the bonding or hydrocarbons which are not themselves soluble in the polar solvent).

If a single washing is employed in the first stage the amount of ethanol or isopropanol which is used should be 1.5 to 3 times the volume of water which is present in the cream or paste of the microorganism. However if desired two washings with ethanol or isopropanol may be employed using in the first washing a volume of solvent equal to the volume of water in the cream or paste and in the second washing a smaller amount of ethanol or isopropanol for example, one half of the amount used in the first washing.

Between washings of each stage or substage the cream or paste is allowed to drain, for example by filtering and part of the residual solvent is then preferably removed by vacuum filtration.

In the second stage the amount of solvent used in the (or each) washing will usually be 2 to 20 times the volume of the resulting dry microorganism.

By the use in the second extraction stage of a solvent which is a mixture of hydrocarbon and a polar solvent the composition of the second stage solvent, which will in any case acquire polar solvent from the first stage extraction, can be stabilized. Build up of polar solvent can be avoided in the course of a distillation stage, in which the second stage solvent is recovered by the removal of separate streams consisting of (a) polar solvent for recycle to the first extraction stage and (b) a mixture of hydrocarbon and polar solvent for recycle to the second stage.

Suitably in a distillation stage the extract obtained by the second extraction stage is distilled to recover (a) overhead a mixture of hydrocarbon, polar solvent and water for recycle to the second extraction stage and (b) a bottoms fraction containing polar solvent, water and the extracted materials; this fraction is preferably blended with the extract obtained in the first extraction stage before this is fed to distillation whereby all contaminants recovered by solvent extraction are removed as a bottoms fraction in this distillation stage. Suitably the polar solvent is ethanol or isopropanol. Suitably the second stage solvent is an azeotropic mixture.

The microorganism fraction, from which as much solvent as possible has been removed by decantation and/or centrifuging and/or filtration, is freed from the remaining traces of solvent and/or water by evaporation, preferably under reduced pressure.

Example: The preparation of the broth in this separation process was conducted as was Example 1 of U.S. Patent 3,536,585 (cf. p. 178). At this point the broth collected at the outlet of the fermenter was subjected to decanting, an aqueous phase consisting of 65% by weight of the broth being withdrawn as spent medium and the remainder of the broth, in the form of a yeast cream recovered. This yeast cream had the composition (percent by weight):

	Percent
Yeast (dry weight)	4.5
Residual gas oil	31.5
Aqueous medium	64.0

Residual gas-oil, apart from that absorbed on yeast, was removed by means of a Sharples supercentrifuge. The yeast after centrifuging had the following composition:

	Percent by Weight
Moisture	68
Lipids (on dry basis)	18
Unsaponifiables (on dry basis) (containing residual adsorbed oil)	40

This recovered yeast was mixed with nutrient medium to achieve a cream of which cellular density was 15 grams/liter.

This yeast cream was continuously fed to a second fermenter identical to the first stage fermenter and operated under conditions identical to the first stage conditions except that no additional substrate or aqueous medium was supplied. Aeration was maintained at 3 millimols of O_2 per liter of medium per minute. Broth was continuously withdrawn and passed to a Sharples supercentrifuge and a yeast paste was recovered. The yeast had, at this stage, the composition:

Moisture	65% by weight (on dry basis)
Lipids	3% by weight
Unsaponifiables	10% by weight
Nitrogen	10% by weight

The water content of the yeast paste was adjusted at a level of 1.5 parts by weight of water per part of dry yeast, yielding a yeast cream. This cream was pumped into an extractor and to it was added an azeotropic mixture of isopropanol and normal hexane at a rate of 8 parts of solvent per part of yeast (dry weight). This extraction step was repeated and these two extraction steps were followed by a second extraction stage consisting of one isopropanol wash at a rate of 8 parts of solvent per part of yeast (dry weight).

Solvents were then drawn off, finally under vacuum. Remaining traces of solvent were stripped, and the yeast dried in superheated steam. The final composition of yeast product was as follows (on dry basis):

	Percent by Weight
Moisture	5
Lipids	0.2
Unsaponifiables	0.1
Nitrogen	11

After these various extraction stages, solvents were recovered, separately or in mixture, and distilled to give:

(a) an azeotropic mixture of isopropanol and normal hexane,
(b) an azeotropic mixture of isopropanol and water; and
(c) a residue fraction.

The recovered solvents were recycled to the extraction stages. The second stage fermenter is seen to constitute a purification stage. In this case no detergent washing was employed to remove absorbed oil.

The process has been further described by *J.A. Filosa; U.S. Patent 3,522,147; July 28, 1970; assigned to The British Petroleum Co., Ltd., England.*

Example: Candida lipolytica was grown in a 12 cubic meter fermenter under continuous conditions in the presence of a culture medium which, as fed to the fermenter, had the composition shown below, the feed rate being 2,400 liters/hour:

	Grams/Liter
$(NH_4)_2HPO_4$	2.0
KCl	1.15
$MgSO_4 \cdot 7H_2O$	0.65
$MnSO_4 \cdot 4H_2O$	0.06
$FeSO_4 \cdot 7H_2O$	0.124
$ZnSO_4 \cdot 7H_2O$	0.306
Yeast extract	0.030
Tap water to 1,000 ml.	

The hydrocarbon feed to the fermenter was a heavy Iraq gas oil containing 12% normal paraffins and having the following characteristics:

Initial boiling point (°C.)	217
5% BP	284
50% BP	343
95% BP	377
Final boiling point	385

The gas oil was fed to the fermenter at the rate of 120 grams per liter of total gas oil and aqueous medium fed. The pH in the fermenter was maintained at 4 by the continuous addition of aqueous ammonia and the fermenter temperature was maintained at 130°C. Aeration was at the rate of 60 vol./vol./hr. of air.

The product of the fermentation was continuously removed and decanted to obtain 800 liters/hour of yeast cream (containing 25 grams/liter of yeast). The cream was diluted with fresh water at the rate of 1,600 liters/hour and mixed vigorously with 0.5 gram/liter of a nonionic surface active agent (obtained by condensing ethylene oxide with a mixture of lauric alcohol and myristic alcohol to give an ethylene oxide chain of an average of 8.5 units per molecule); the surface active agent is marketed as NI 29.

The mixture was centrifuged in a Sharples DG2 autojector type centrifuge to recover the following three phases: yeast cream, residual gas oil, aqueous phase.

The yeast cream was mixed with fresh water to which had been added 10% of its volume of the culture medium as used in the fermenter and the mixture was continuously fed to a vessel in which the residence time was 3 hours while maintaining gentle aeration at the rate of 10 vol./vol./hour of air.

The product of this stage was centrifuged in a simple nozzle-separator Alfa-Laval QX centrifuge. The yeast paste was drum dried at a temperature of 140°C. and the yeast product was solvent extracted with normal hexane at a rate of 2 liters of normal hexane per kilogram of yeast (measured as dry yeast) present in the yeast products; the solvent extraction was repeated 5 times to give a substantially hydrocarbon-free yeast product.

Solvent Extraction

B.M. Laine, J.C. Hondermarck and R. Goux; U.S. Patent 3,616,209; October 26, 1971; assigned to The British Petroleum Company Limited, England have described the solvent extraction of solid material which is contaminated with at least one hydrocarbon and water.

The method includes a first extraction stage consisting of one or more extraction steps, extracting a contaminated solid material with a mixture of an alcohol and a hydrocarbon with which it forms an azeotrope, the alcohol and azeotrope forming hydrocarbon respectively being employed at a ratio by volume within the range 30:70 to 70:30.

In a second extraction stage consisting of one or more extraction steps, following the first extraction stage, the treated solid material from the first stage is extracted with a solvent comprising or consisting of the alcohol, thereafter recovering the treated solid material.

Suitably the azeotrope-forming hydrocarbon is normal hexane. The alcohol is ethanol, propanol, isopropanol or a butanol. The process is applied to a crude or partially refined product of the growth of a microorganism on a hydrocarbon substrate in the presence of an aqueous nutrient medium. Preferably the microorganism is a yeast.

The extract fractions recovered from the extraction stages are fed, separately or after blending, to a distillation stage, consisting of one or more distillation steps, for the separate recovery of (a) an azeotropic mixture of the alcohol and the azeotrope-forming hydrocarbon,

(b) an azeotropic mixture of the alcohol and water and (c) a residue fraction, thereafter blending part of the azeotropic mixture of the alcohol and water with part of the azeotropic mixture of the alcohol and azeotrope-forming hydrocarbon, the parts being selected to give a mixture of the alcohol and azeotrope-forming hydrocarbon containing these materials respectively at a ratio by volume in the range 30:70 to 70:30 and recycling this mixture to the first extraction stage.

Example: 10 liters of the following aqueous mineral medium was introduced into a 15-liter stirred fermenter; parts are by weight.

Sodium phosphate tribasic	3.4
Potassium chloride	0.6
Magnesium sulfate	0.3
Ammonium sulfate	2.5

Made up to 1,000 parts with soft water containing trace elements. A suitable alternative medium has the composition:

Diammonium phosphate	2
Potassium chloride	1.15
Magnesium sulfate · $7H_2O$	0.65
Zinc sulfate	0.17
Manganese sulfate · H_2O	0.045
Ferrous sulfate · $7H_2O$	0.068
Tap water	200
Yeast extract	0.025
Distilled water (to make up to 1,000 parts).	

To the fermenter was added a few parts per million of yeast extract and then 50 grams of *Candida lipolytica* in the form of an aqueous cream containing 20% by weight of dry material and then 150 grams of a heavy gas oil of petroleum origin containing 20% by weight of normal paraffins.

When the culture reached the desired concentration of yeast cells for continuous operation, the continuous feed, to the fermenter, of aqueous mineral medium and petroleum oil was started up. The temperature was maintained at 30°C. and the pH of the medium was maintained regulated at pH 4 by the addition of aqueous ammonia.

This emulsion was fed to a centrifugal separator from which were recovered three phases, being, in increasing order of density; (a) an oil phase containing the yeast cells, (b) an aqueous mineral medium phase (which may contain traces of oil and yeast) and (c) a yeast cream containing approximately one part of yeast, four parts aqueous medium and a certain quantity of oil adhering to the yeast cells. The yeast cream and an aqueous solution of a surface-active agent, were fed continuously to a mixer.

The surface-active agent was used at an aqueous concentration of 0.05% by volume, two parts by volume of the aqueous solution being added to one part by volume of the yeast cream. The surface-active agent was a material sold as NI 29 and being the product obtained by condensing a mixture of lauric alcohol and myristic alcohol with ethylene oxide, the product having an ethylene oxide chain of an average 8.5 units per terminal group.

The mixture so obtained was centrifuged to obtain three fractions in increasing order of density: (a) an oil phase, (b) an aqueous phase containing the surfactant product which was recycled to the mixer and (c) a second yeast cream containing one part by weight of yeast (which was still slightly contaminated by the oil) with four parts by weight of aqueous liquid containing the surfactant.

This second yeast cream was passed with water to a mixer and the mixture so obtained was centrifuged to obtain: (a) an oil phase, (b) an aqueous phase and (c) a thick yeast cream containing 20% by weight of yeast (estimated as dry yeast) and 80% by weight of water and which contained only a very small quantity of oil.

After a subsequent water-rinsing followed by centrifuging there was obtained a yeast product containing 65% by weight of water, together trace amounts of hydrocarbon contaminant. By further centrifuging there was obtained an impure yeast cream containing circa one part by weight of dry contaminated yeast per 1.5 parts by weight of water.

This yeast cream was then pumped into an extractor which was in the form of a filtering drum which was rotated with its axis horizontal. A solvent mixture consisting of three parts in volume of isopropyl alcohol and five parts in volume of an azeotropic mixture (of which the composition is normal hexane 80% by weight and isopropyl alcohol 20% by weight) was poured into the extractor, the volumes of solvent being based on one part by volume of dry (contaminated) yeast.

The mixture was maintained at 80°C. for 20 minutes and then the solvent drawn off, finally under vacuum. A second extraction step was carried out identical to the first.

In a second extraction stage the solid material which was recovered was washed with isopropyl alcohol at a rate of 4 volumes for 1 volume of dry yeast at 50°C. for 90 minutes. The remaining solvents and contaminant were removed. Finally the yeast product was dried in superheated steam.

The analysis of the former yeast cream which was treated by solvent extraction and the final yeast product after solvent extraction is given in the following table.

	Yeast Cream	Final Yeast Product
Nitrogen % by weight of dry yeast	9.8	11
Total lipids % by weight of dry yeast	13	0.5

Yeast product thus obtained, freed of contaminant, can be used as a fodder material.

Use of a Drum Dryer

A.J. Clark and R.J. Hunt; U.S. Patent 3,595,749; July 27, 1971; assigned to The British Petroleum Company Limited, England have developed a process for the removal of contaminants from a mixture of a microorganism and contaminants, the contaminants consisting of or including a hydrocarbon oil and water.

The procedure involves subjecting the mixture to drum drying in a dryer having a pair of hot rolls which constitute a nip. At least part of the water is removed by evaporation and an oil enriched product is removed from above the nip and a microorganism enriched product is removed, from at least one roll, below the nip.

When operating with a drum dryer, the microorganism mixture which passes the nip has a favorable character even though the mixture fed into the nip is of a pasty or tacky character. The change in character is believed to be caused by a redistribution of the cells of the microorganism in the mixture under the shearing forces of the nip which causes the mixture, which is a dilatent body, to dry out superficially.

In consequence the mixture immediately after passing the nip can be removed from the drum by knifing, this notwithstanding that the mixture subsequently relaxes and again takes on a tacky character.

Whether or not the mixture after removal from the roll becomes tacky it is found that this mixture retains to some degree a flaky character which is a beneficial quality when the mixture is subsequently solvent extracted and particularly when the solvent diluted mixture is subjected to filtration for the removal of the solid microorganism phase.

As a result of the treatment in the nip of a drum dryer, the filtration stage or stages can be operated more efficiently with good throughput rates.

Example 1: 10 liters of the following aqueous mineral medium was introduced into a 15 liter stirred fermenter; parts are by weight:

Sodium phosphate, tribasic	3.4
Potassium chloride	0.6
Magnesium sulfate	0.3
Ammonium sulfate	2.5

Made up to 1,000 parts with soft water containing trace elements. To the fermenter was added a few parts per million of yeast extract and then 50 grams of *Candida tropicalis* in the form of an aqueous cream containing 20% by weight of dry material and then 150 grams of a heavy gas oil of petroleum origin containing 20% by weight of normal paraffins.

When the culture reached the desired concentration of yeast cells for continuous operation, the continuous feed, to the fermenter, of aqueous mineral medium and petroleum oil was started up. The temperature was maintained at 30°C. and the pH of the medium was maintained at pH = 4 by the addition of aqueous ammonia.

This emulsion was fed to a centrifugal separator from which were recovered three phases being, in increasing order of density; (a) an oil phase containing trace amounts of yeast cells, (b) an aqueous mineral medium phase (which may contain traces of oil and yeast) and (c) a yeast cream containing approximately 1 part of yeast, 4 parts of aqueous medium and a certain quantity of oil adhering to the yeast cells.

The yeast cream was fed to a drum dryer consisting of twin horizontal rollers providing a nip and having the following characteristics:

Diameter of rolls	18 inches
Width of rolls	18 inches
Temperature	140°C.
Rate of revolution	30 revolutions per minute

Gas oil collected in the nip of the rolls and was drawn off. Dry yeast having a reduced content of gas oil (that is, having a gas oil content much less than the content of gas oil per unit weight of yeast in the feed material) was withdrawn by means of scrapers from below the nip of the rolls.

Example 2: The material treated was a mixture of *Candida tropicalis,* water and gas oil. This material was fed to a drum dryer of the type and under the conditions described in Example 1. The composition of the feed material and of the yeast product are shown in Table 1.

TABLE 1

	Percent Weight		
	Solids	Oil	Water
Feed material	21	20	59
Drum dried yeast	75	20	5

Thus the amount of gas oil based on gas oil and solids was reduced from 48% in the feed material to 21% in the dried product.

Example 3: The material treated was a mixture of bakers' yeast (*Saccharomyces cerevisiae),* water and gas oil. This material was fed to a drum dryer of the type and under the conditions described in Example 1. Composition of feed material and of the yeast product are shown in the following Table 2.

TABLE 2

	Percent Weight		
	Solids	Oil	Water
Feed material	28.0	17.0	55.0
Drum dried yeast	70.8	18.8	10.4

Solvent Removal by Steam

I.C. Bennett and J.A. du Chaffaut; U.S. Patent 3,701,713; Oct. 31, 1972; assigned to The British Petroleum Company Limited, England have described a method for removing solvent from a microorganism containing fermentation product which has been produced by cultivating a microorganism in the presence of a hydrocarbon and subjecting the cultivated microorganism to a solvent extraction treatment to remove residual hydrocarbon.

The method consisting essentially of passing steam over the heated fermentation product in such a manner as to maintain in the product a certain minimum water content while the solvent is being removed.

In particular in fermentation processes for the cultivation of microorganisms on hydrocarbons as a consumable carbon source the cultivated microorganism-containing biomass may be treated with solvents to remove traces of residual hydrocarbons. The solvent, usually consisting of or containing an alcohol, can be removed from the biomass by evaporation for example under reduced pressure.

It has been found that removal of the solvent can be improved by maintaining a certain minimum quantity of water in association with the microorganism during the solvent removal treatment, namely at least 20% by weight of water in relation to the weight of microorganism in the "dry state", product obtained by drying at 120°C.

According to this method there is provided a process for the removal of solvent from fermentation products consisting essentially of microorganisms which process comprises contacting a biomass, containing a solvent consisting of or containing an alcohol, with moving heated surfaces while continuously feeding biomass into contact with the surfaces, moving biomass transversely with respect to the heated surfaces and withdrawing biomass from contact with the surfaces.

The biomass is maintained in an atmosphere consisting of a vapor phase consisting of or containing steam with vapor of the alcohol; steam being introduced continuously, the vapor phase moving cocurrently or countercurrently to the moving biomass, the rate of introduction of steam being at least sufficient to ensure that, under the conditions of drying employed, substantially all of the solvent is removed from the biomass being treated.

The biomass is contained in a horizontal trough and is contacted with heated rotating discs mounted on a rotated horizontal shaft. The discs are hollow and are heated by a heating medium circulated through the shaft. The alcohol is ethanol or isopropanol.

The biomass may be obtained by continuously cultivating a microorganism in the presence of an aqueous nutrient medium, a hydrocarbon substrate and a free oxygen-containing gas, the microorganism being of a strain which consumes the hydrocarbon, recovering a biomass consisting of or containing the microorganism contaminated with hydrocarbon, solvent extracting the biomass with a solvent consisting of or containing an alcohol, recovering a biomass containing the solvent.

The hydrocarbon substrate consists at least in part of a straight chain hydrocarbon. Usually the straight chain hydrocarbons will be those present in petroleum feedstocks as paraffins; however, the straight chain hydrocarbons may be present as olefins; also there may be used a mixture containing straight chain paraffins and olefins.

Suitable petroleum feedstocks include kerosene, gas oils and lubricating oils; these feedstocks may be unrefined or may have undergone some refinery treatment, but must contain a proportion of straight chain hydrocarbons in order to fulfil the purpose of this process. Suitably the petroleum fraction will contain 3 to 4% by weight of straight chain hydrocarbons.

The process is further described in the accompanying drawing which shows a longitudinal section of an apparatus for removing solvent from contaminated biomass.

The apparatus consists of a horizontal cylindrical drum **1** having a length of 1.25 meters and an internal diameter of 0.5 meter. The drum has an external jacket **2** having a heating area of one square meter into which steam or other heating medium may be passed. A hollow rotatable shaft is axially located within the drum. This shaft **3** is equipped with spaced hollow discs **4** mounted in a plane normal to the shaft axis.

Steam or other heating medium can be passed into the hollow shaft and discs which together provide a heating area of 4.5 square meters. Scrapers **5** are fitted at intervals on the periphery of each disc to prevent treated material from caking on the internal surface of the drum. The top internal surface of the drum is provided with a

FIGURE 5.2: SOLVENT REMOVAL FROM MICROORGANISMS

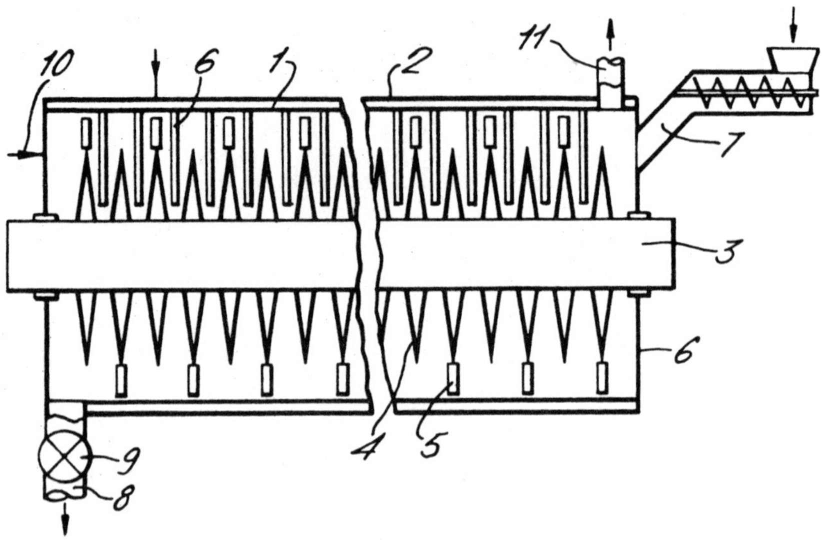

Source: I.C. Bennett and J.A. du Chaffaut; U.S. Patent 3,701,713; October 31, 1972

plurality of scraper blades **6** which project radially into the drum between the discs to prevent treated material from caking on the discs. An inlet **7** through which the material to be treated can be passed into the drum is provided near the top and at one end of the drum.

An outlet **8** through which the treated material is passed out of the drum, is provided near the opposite lower corner of the drum. This outlet is controlled by rotary valve **9** which regulates the rate at which treated material leaves the drum.

A steam inlet pipe **10** (schematically shown) is fitted at the top of the drum at the end opposite to the end having the inlet and a steam/vapor exit pipe **11** is located near the inlet. If desired the top surface of the drum and heating jacket may be recessed with a chamber having substantially rectangular cross section to provide a zone for the disengagement of steam/vapor from the solid treated material.

In operation of the apparatus 77.8 kilograms per hour of a biomass consisting of 45% by weight of dry matter, 43% by weight of isopropyl alcohol and 12% by weight of water was fed at a temperature of 75°C. into the drum through inlet **7**. The biomass was obtained by cultivating a hydrocarbon consuming yeast on a hydrocarbon and aqueous nutrient medium; separating the yeast from the growth medium, and treating it with isopropanol to remove residual hydrocarbons.

The feed rate was equivalent to 35 kilograms per hour of dry matter in the biomass. Saturated steam at a pressure of 1 kilogram per square centimeter gauge was supplied to the heating jacket, shaft and discs. The shaft was rotated at a speed of 30 rpm.

Saturated steam at substantially atmospheric pressure was fed into the body of the drum through pipe **10** at the rate of 35 kilograms per hour. The treated biomass withdrawn from outlet **8** contained 0.4% by weight of alcohol and 22% by weight of water in relation to the weight of dry matter in the biomass.

RECOVERY OF PROTEINS

Enzymatic Rupture of Cell Walls

A. Champagnat and B.M. Laine; U.S. Patent 3,268,412; August 23, 1966; assigned to The British Petroleum Company Limited, England have provided a process which comprises cultivating a microorganism in the presence of a feedstock containing straight chain hydrocarbons; and thereafter subjecting a fraction containing microorganism to conditions such that proteins or polyproteins or polypeptides or amino-acids are recovered.

The fraction containing the microorganism is subjected to autolysis, plasmolysis or hydrolysis. Autolysis may be brought about by means of the enzymes contained in the microorganism or by the addition of other enzymes.

The cultivation of the microorganisms, for example yeasts, on petroleum hydrocarbons gives cells which, after washing and drying, have a particularly sharp and rancid taste. These microorganisms, particularly yeasts, have a high content of lipids, the presence of which is undesirable because of their easy oxidation which leads to a disagreeable taste and odor.

However, the dry microorganism even when freed from lipids have an insufficient and irregular digestive utilization rate. This disadvantage is at least in part due to the condition of drying the cells which is necessarily somewhat severe. The process converts the microorganism to a more digestible material.

Autolysis, which is the favored method for breaking down the cells, leads to a product in the form of a liquid, the viscosity of which depends on the concentration of dry matter. The autolyzate which contains proteins of a much lower molecular weight than the dry yeast is much more digestible. Since it has been obtained without severe heating, the unstable amino acids and the vitamins may be completely or substantially preserved.

Furthermore, the liquid form of the autolyzate permits of an easy extraction of the lipids which it contains, using an appropriate solvent.

Thus the autolyzate may be treated with a suitable solvent to extract from it the lipids and the traces of hydrocarbons which it may still contain. It is then preferably concentrated in vacuo at a temperature lower than 70°C. so as to preserve from destruction the desired constituents, for example, amino acids and vitamins.

A microorganism, preferably in the form of a cream, is subjected to treatment for the separation of proteins or polyproteins or polypeptides or amino acids contained in the microorganism from cell wall material.

Autolysis is carried out at a temperature in the range 45° to 55°C., preferably at 50°C. The autolyzate is usually obtained in the form of a liquid the viscosity of which depends upon the concentration of dry substances which it contains. As it is obtained without violent heating the amino acids which are not very stable and the vitamins are preserved, usually completely.

When an autolyzate is prepared for cattle feeding, a yeast is autolyzed for three hours. Preferably this autolyzate will contain 40 to 50% by weight of dry matter.

When an autolyzate is prepared for human consumption a yeast is autolyzed for 10 to 20 hours. This autolyzate will contain 55 to 65% by weight, preferably 60% by weight of dry matter.

According to another aspect there are manufactured instead of autolyzates, hydrolyzates obtained by the action of acids or alkalies on the cells of microorganisms or plasmolyzates by the diffusion of the proteins out of the cells as a result of the action of a change in osmotic pressure.

This change is achieved, for example, by the addition of sodium chloride to the aqueous medium. The operations of autolysis, hydrolysis and plasmolysis may be carried out in the same type of apparatus. Preferably this apparatus is a stirred vat.

By reducing the hydrocarbon content of the microorganism by washing, the rate of operation of the treatment for the rupture or extraction of the cell walls, for example, by autolysis, is increased; in consequence the residence time of the microorganism in this stage will out-weigh the costs of the washing stages.

The product of this stage for the rupture of the cell walls is treated for the separation of the nitrogenous materials. This separation may be carried out by centrifuging and/or filtration. The nitrogenous fraction so obtained will contain as impurities hydrocarbons and usually also lipids. This fraction will preferably be subjected to solvent extraction. However, if desired solvent extraction may be carried out prior to or without separation of the cell wall material.

Example: A culture product was centrifuged and a yeast paste was recovered. 8.4 kilograms of this paste contained 2 kilograms of dry material (excluding retained hydrocarbons), 1 kilogram of nonmetabolizable hydrocarbons and 1 kilogram of proteins; to this paste was added 14.5 liters of water containing 180 grams of sodium chloride.

The mixture was maintained at 50°C. for 10 hours under continuous slow agitation; without controlling the pH. The product so obtained contained about 600 grams of protein and 1,200 grams of solubilized dry material.

This product was centrifuged in two stages in a Sharples centrifuge at 11,000 *g* to recover an oil phase containing cell wall material; a liquid autolyzate (containing 500 grams of protein) and a residue containing cell wall material, slightly impregnated with hydrocarbons.

The cell wall material recovered as residue was subjected to partial drying. After drying to a product containing 20% by weight of water, this product was deoiled by extraction at 55°C. by means of an azeotropic mixture of normal hexane and isopropanol at a ratio of 5 liters of solvent per kilogram of dry material in the product treated. The deoiled material can be used as an animal foodstuff.

Heating in an Aqueous Acidic Solution

Bacterial cells can be mass-produced industrially from carbon sources such as sugars and hydrocarbons, simple nitrogen sources such as ammonia and ureas and from inorganic salts. Hope has been held that such cells would provide sources of protein.

Generally, microorganisms possess mechanically and chemically strong cell walls and when used for foodstuffs as they are, the digestion ratio thereof is unfavorable. Moreover, the taste is not always satisfactory depending upon the particular type of microorganism. Development of a process for extracting proteins from microorganisms has a great significance.

Processes have been developed for extracting proteins from microorganisms which involve recovering proteins extra-cellularly which have been liberated after mechanical grinding and supersonic treatment of bacterial cells of microorganisms.

However, such processes require a great amount of energy for breaking of the bacterial cells. Processes have also been reported where proteins are extracted after treating the bacterial cells with solvents and surface active agent containing solutions. But, these processes are less effective for certain kinds of microorganisms, especially with gram-positive bacteria.

H. Samejima, H. Teranishi and T. Deguchi; U.S. Patent 3,585,179; June 15, 1971; assigned to Kyowa Hakko Kogyo Co., Ltd., Japan have found that the majority of proteins contained in bacterial cells can be easily extracted in favorable yields by preliminarily heating the cells with dilute acids under comparatively moderate conditions and then extracting with water, salt solutions such as sodium chloride, alkali solutions such as sodium hydroxide, calcium hydroxide, potassium hydroxide and ammonium hydroxide, and/or urea solutions.

The extraction may be effected at any suitable pH in the vicinity of neutral (but, in the case of urea solutions, at neutral or slightly alkaline pH's). The process may be practiced easily, cheaply and efficiently from the industrial standpoint on microorganisms generally known. For example, a urea extraction achieved under neutral or weakly alkaline conditions, after subjecting the microorganism cells preliminarily to proper heat treatment in dilute acids, makes it possible to extract proteins of good quality in more favorable yields than that obtained by the individual use of the urea extraction method per se.

It is believed that the acid treatment ruptures a portion of the bonds in the polysaccharides, thereby destroying a part of the cell walls and facilitating the extraction of proteins from the cells.

Protein extraction from the cell walls of gram-positive bacteria with alkaline solutions is generally difficult because of the high content of polysaccharides, but according to the process, the proteins can be easily extracted.

The process may be carried out, for example, as follows. Protein extraction is achieved comparatively easily and efficiently by suspending living or dry bacterial cells of microorganisms in dilute acids, for example, having normalities ranging from 0.5 to 2, and then treating the suspension of cells with heat for a short time at any temperature above room temperature, favorably at 60° to 100°C.

The suspension, or the bacterial cells recovered therefrom, is then dipped in water or extracting agent at any pH in the vicinity of neutral (except in the case of urea solutions, wherein a pH of around neutral or slightly alkaline is employed, i.e., from about 6 to 9). The latter treatment may be accomplished at room temperature or with heating.

It has been considered that the acid treatment of proteins causes denaturation and decomposition, and such a treatment has never been used as a pretreatment for the extraction of proteins from the cells of microorganisms.

The table which follows shows the extraction ratios of total nitrogen and protein nitrogen obtained by extracting proteins from *Brevibacterium ammoniagenes* ATCC 6872, a gram-positive bacterial strain, with water and aqueous caustic soda (sodium hydroxide) solutions of various concentrations, alternatively by extraction with aqueous caustic soda solutions after the preliminary heating of the microorganisms in dilute hydrochloric acid.

In this experiment 10 grams of living microorganisms of *Brevibacterium*

Nitrogen Extraction Ratios from *Brevibacterium ammoniagenes*

Condition for pre-treatment				Condition for extraction				Extraction ratio of nitrogen fraction, percent	
Concentration of hydrochloric acid	Amount of hydrochloric acid added, ml.	Treating temperature, C.	Treating time, hours	Extracting agent	Amount of extracting agent added, ml.	Treating temperature, ° C.	Treating time, hours	Extraction ratio of total nitrogen [1]	Extraction ratio of protein nitrogen [2]
				Water	20	24	3	11.5	3.2
				0.5% NaOH	20	24	3	8.3	2.7
				1% NaOH	20	24	3	9.0	3.5
				2% NaOH	20	24	3	13.7	4.0
				4% NaOH	20	24	3	16.2	4.5
				8% NaOH	20	24	3	18.2	5.0
				2% NaOH	40	24	3	18.8	8.0
				4% NaOH	40	24	3	20.6	8.5
				2% NaOH	40	40	3	27.3	13.3
				do	40	60	3	28.0	14.5
1 N	20	98	1	do	[3] 40	24	3	41.7	32.6
2 N	20	98	1	do	[3] 40	24	3	74.5	42.7

[1] Calculated by presuming that total nitrogen in used bacterial cells was 100%.

[2] Calculated by presuming that protein nitrogen in used bacterial cells was 100%.

[3] NaOH was added so that the NaOH was 2% of the total when the volume of solution was brought to 40 ml., since 20 ml. of acid had been preliminarily added.

ammoniagenes [water content 62.2%, total nitrogen content 9.6% (to dry weight) and protein nitrogen 8.3% to dry weight] were used in each case.

As shown in the table, the extraction of a living microorganism such as *Brevibacterium ammoniagenes* with water or aqueous caustic soda solutions of various concentrations gives extremely low extraction ratios of proteins.

On the other hand, those preliminarily treated with 1 to 2 N hydrochloric acid at 98°C. for one hour show extremely high extraction ratios both of total nitrogen and of protein nitrogen (nitrogen insoluble to heated 7% trichloroacetic acid) after subsequent extraction with caustic soda.

Moreover, the cells subjected to a preliminary acid treatment show far higher protein extraction ratios than that observed when a higher temperature condition (for example, 40° and 60°C.) is used for untreated materials in an extraction step with alkali.

Example 1: Five hundred grams of dried cells of *Corynebacterium glutamicum* (*Micrococcus glutamicus*) ATCC 13761 (total nitrogen content, 14.64%; protein nitrogen content, 9.1%) are suspended in 25 liters of 1 N hydrochloric acid and heated for one hour in a boiling water bath with stirring, to which 25 liters of 2 N caustic soda aqueous solution is added after cooling. Protein is extracted by continuation of stirring for 3 hours at room temperature.

The extract and residue are subsequently separated by centrifugation, and the extract (5.65 liters) is adjusted to a pH of 4.0 by the addition of 6 N hydrochloric acid and is allowed to stand overnight in a cold room regulated at 5°C. to complete the precipitation and sedimentation of protein.

The protein fraction precipitated is collected by centrifugation, and an adequate amount of water is added thereto to make a porridge-like fluid which is dried into a powder by a spray dryer. The yield is 368 grams, which represents 73.6% of the weight of original microorganisms.

The powdered protein thus obtained consists of 6.4% water, 70.3% of crude protein content (total nitrogen percent times 6.25) and 2.3% of reducing sugar (as glucose). It uniformly disperses in water at a pH of 7.0. An artificial digestion ratio test by trypsin shows a digestion ratio of 78.7%, which indicates that the protein is of a very good quality as compared with a commercial protein extracted from soy beans, the latter showing a digestion ratio of 50.6% as determined by the same method.

Example 2: One hundred grams of dried cells of *Corynebacterium glutamicum* (*Micrococcus glutamicus*) ATCC 13761 total nitrogen content, 12.2%; protein nitrogen content, 9.1% are suspended in 500 ml. of 1 N hydrochloric acid, heated in a boiling water bath with stirring, kept for 30 minutes after the internal temperature rises up to 95°C., whereby the pH of the suspension is adjusted to 7.0 by the gradual addition of 10 N sodium hydroxide, and extracted by the addition of 1 liter of a 10 M urea solution at 60°C. for 24 hours with stirring.

The extract is centrifuged to remove solid constituents. The solid portion is further washed with 200 ml. of water, which is then centrifuged and combined with the above supernatant. The combined solution is dialyzed in 6 dialyzing membrane bags, about 300 ml. in each, against running tap water.

After 72 hours of dialysis, the contents of the bags are combined, adjusted to a pH of 4.2 with hydrochloric acid and allowed to stand overnight in a cold room at 5°C. The resulting precipitate is separated from the solution by centrifugation. As a result, 40.2 grams of lyophilized powder is obtained.

The water content and protein content (total nitrogen percent times 6.25) of the resultant powder are 6.5% and 85.2%, respectively. This protein is uniformly dispersed in water when adjusted to a pH of 7. An artificial digestion ratio observed with trypsin is 90.6%, showing that the product is a protein of very high quality as compared with a commercial protein extracted from soy beans, the latter showing a digestion ratio of 50.6% as determined by the same method.

Acid Treatment with Perchloric and Trichloroacetic Acids

J.W. Frankenfeld and B.L. Dasinger; U.S. Patent 3,634,194; Jan. 11, 1972; assigned to Esso Research and Engineering Company has described a fermentation process for the production of high-quality protein cells from petroleum hydrocarbons and oxygenated hydrocarbons using a particular technique for the coagulation of the produced cells. Very desirable acids for use in the process are selected from the group consisting of perchloric acid and trichloroacetic acid.

These acids are unique in that they serve the dual purpose of coagulating the cells and removing the nucleic acids contained therein.

Hydrocarbon fractions, such as petroleum liquids, are used as such or are oxidized, or partially oxidized, and then contacted with a microorganism under fermentation conditions to produce high yields of protein.

The fermentation process may be carried out continuously or batchwise. The fermentation broth from the fermentation stage or initial stage containing the cells is introduced into a coagulation stage wherein the cells are coagulated using a particular acid wherein the nucleic acids are separated from the coagulated cells and are in the supernatant phase. This phase is then withdrawn and handled to recover the nucleic acids which are useful, for example, as flavoring agents and other industrial applications. Also, by removing these nucleic acids the protein value of the cells is enhanced.

The biosynthetic process is applicable to the biosynthesis of all microorganisms which are capable of growth on petroleum hydrocarbon and oxygenated hydrocarbon substrates, particularly those oxygenated substrates derived from petroleum hydrocarbon fractions.

Acid is sufficient to lower the pH of the broth to from about 2 to 4.5 such as about 3.5. By using the acids to secure the pH specified, a marked improvement in accelerating product recovery is secured. The microbial cells are rapidly coagulated prior to separating them from the product containing effluent stream.

The rate at which the microbial cells are separated from the remainder of the product stream is accelerated by as much as 20-fold or even higher, thus attaining substantial economies in time and equipment necessary for conducting the separation procedure on a given volume of product stream.

After addition of the acid, the acidified effluent is preferably heated. The heating is conducted at a temperature above the growth temperature maintained in the biosynthesis bath yet below the temperature at which cell degradation occurs (protein degradation).

Temperatures can be used which cause protein denaturation as long as no degradation of food value occurs. The heating is continued for a sufficient period of time to attain substantial coagulation of cells and removal of nucleic acids.

Example: A fermentation process was conducted using a petroleum hydrocarbon feed containing from about 12 to 20 carbon atoms. The temperature in the fermentation balance was about 35°C. and the pH about 7.0. The microorganism utilized was *Micrococcus cerificans.*

Three 5-cc portions of the broth were removed from the fermentation zone. The cells from each sample were removed by centrifugation, washed with water and resuspended in 5 cc of distilled water. The three samples were treated as follows:

Sample 1 was not further treated and was used as a control.

Sample 2 was acidified to pH 4.0 by addition of HCl. The volume was adjusted to 10.0 cc and the sample was heated at 65°C. for 10 minutes. After cooling the cells were removed by centrifugation and the ultraviolet absorbance of the supernatant measured at 260 mμ, the typical wavelength for nucleic acids. The absorbance at this wavelength is a measure of the amount of nucleic acids extracted into the supernatant.

Sample 3 was treated in the same manner as Sample 2 except that 5.0 cc of 0.5 N PCA was added in place of the HCL. In addition, both the supernatant of Sample 3 and the residual cells after removal of the supernatant were analyzed for protein content. This was done in order to determine the amount of protein lost by PCA extraction.

It is apparent that treatment with PCA and heating is very effective in removing nucleic acids while addition of HCl plus heating is not. In Sample 3 (PCA + heat) 95% of the nucleic acids were removed while in Sample 2 (HCL + heat) only 13% were extracted. In addition, during PCA treatment only 2.4% of the cell protein was lost. The effectiveness of perchloric acid in removing nucleic acid from cells is shown in the table below.

		Nucleic Acids Removed		
Sample	Treatment	μg./mg. Cells	%	Protein Lost %
1	None	2.3		nd*
2	HCl + Heat	15.8	13	nd*
3	PCA + Heat	118.0	95	2.4

*nd = not determined.

Solvent Extraction of Lipids

A. Champagnat and B. Laine; U.S. Patent 3,268,419; August 23, 1966; assigned to The British Petroleum Company Limited, England have described a process for the production on hydrocarbons of edible microorganisms in purified form, for example, yeast, bacteria of the orders of Pseudomonadales, Eubacteriales, Actinomycetales and *Penicillium expansum.*

The cultivation of yeasts on hydrocarbons, and in particular on petroleum hydrocarbons, as the carbonaceous substratum, generally gives yeasts which after washing and drying possess a peculiar taste which is sharp and rancid and which may interfere with their use in foods for human consumption.

The traditional yeasts, cultivated on molasses, residuary liquors or bisulfite liquors, also have a peculiar taste, different from that of yeasts cultivated on hydrocarbons, and certain processes have been proposed for attenuating this taste, notably to attenuate their bitterness, but none of these has succeeded completely.

It has been found that yeasts cultivated with hydrocarbons as the sole source of carbon have a distinctly higher lipid content than yeasts cultivated on the traditional substrate: molasses, residuary liquors, bisulfite liquors. Extraction of these lipids, wholly or in part, leads to a substantial reduction in this characteristic taste of yeasts cultivated on hydrocarbons or to the elimination of this taste.

The process involves cultivating a microorganism in the presence of a feedstock consisting wholly or in part of straight chain hydrocarbons, separating from the product a fraction comprising the microorganism and extracting the fraction by means of a solvent.

Solvents which may be employed include ethyl alcohol, isopropanol, light hydrocarbons, including benzene and light platformate fractions, ethyl ether, acetone, chlorinated solvents and liquefied petroleum gases, such as butane and propane.

If desired a first extraction stage can be operated using a polar solvent, for example an alcoholic solvent, ethanol or isopropanol and then the partially purified microorganism can be further treated in a second extraction stage using a hydrocarbon solvent, for example normal hexane or a light platformate fraction or benzene. Preferably in the second stage there is used as solvent a mixture of hydrocarbon in major amount with a polar solvent in minor amount such as an azeotropic mixture of hexane with isopropanol or ethanol. If desired both extraction stages can be operated in continuous manner.

By the use of any alcohol in the first stage of a two-stage extraction, the water content of the microorganism cream or paste is considerably reduced. As a result the microorganism containing material which is fed to the second stage has a sufficiently low content of water to ensure that the nonaqueous contaminants which are still present are miscible with the solvent used in the second extraction stage.

Each extraction stage may consist of either one or more substages consisting of washing with the solvent used in the stage followed by separation.

Example 1: A yeast cream of the strain *Candida lipolytica* was prepared. In a series of runs a solvent was pumped continuously into an extractor containing a batch of the yeast cream. The extractor was in the form of a filtration drum which was rotated with its axis horizontal.

Yeast cream containing 2 parts by weight of dry yeast and 8 parts by weight of water was charged to the extractor with 16 parts of ethanol. The mixture was maintained at 60°C. for 30 minutes while rotating the drum.

Solvent was drawn off, finally under vacuum. There was thus obtained a mixture of 2 parts of yeast and 2 parts of solvent which consisted of ⅓ water and ⅔ of ethanol, together with some remaining contaminants. This mixture was treated with 10 parts by weight of a solvent mixture consisting of: 80% by weight normal hexane and 20% by weight ethanol.

The mixture was maintained at 60°C. for 15 minutes and solvent drawn off, finally under vacuum. The washing with this solvent mixture was repeated several times. If a yeast product of high purity is required the washing stage may be repeated up to, for example, 10 times.

Finally the yeast product is dried in super heated steam. The analysis of the yeast cream before solvent extraction and of the yeast product obtained is shown in the following table.

	Yeast Cream	Yeast Product
Nitrogen, by weight of dry yeast	7%	7.8%
Total weight of lipids based on weight of dry yeast	10%	0.5%

Example 2: The yeast cream described in Example 1 was continuously extracted in a drum having its axis vertical and having a paddle stirrer which rotated at 10 rpm. In separate runs the solvent was respectively ethanol; ethanol followed by normal hexane; and isopropanol.

The extract was continuously distilled for the recovery of lipids and hydrocarbons and the solvent recycled to the extractor. Conditions and results obtained were as shown in the following table.

Run No.	4		7		8	
Solvent and Period of Treatment	Ethanol (11 hrs.)		Ethanol (20 hrs.) Hexane (9 hrs.)		Isopropanol (40 hrs.)	
Rate of Feed of Solvent	600 cc/hr.		600 cc/hr.		600 cc/hr.	
Temperature	60°C.		60°C.		60°C.	
Weight of Yeast Paste (grams)	1,000		1,000		1,000	
Analysis	Before Treatment	After Treatment	Before Treatment	After Treatment	Before Treatment	After Treatment
Dry Yeast:						
Wt. (grams)	210	140	210	200	210	200
% Nitrogen	6.05	9.0	5.65	7.45	5.65	8.50
Total % wt. of Lipids in Yeast	41.2	5.0	42.9	4.2	42.9	6.9
Extract, wt. (grams)	-	72	-	200	-	100

Precipitating Extracted Protein at Acid pH

J.W. Frankenfeld and C.J. McCoy; U.S. Patent 3,347,688; Oct. 17, 1967; assigned to Esso Research and Engineering Company have been able to obtain high protein containing adhesives from microorganism cells capable of growing on hydrocarbon feeds. High bacterial protein-containing adhesives and paper coating binders are obtained from *Micrococcus cerificans* cells by lysing the cells, conducting a basic extraction on the lysed cells and isoelectrically precipitating the extracted protein by adjusting the pH in acidic medium. The precipitated protein curd is collected by centrifugation, washed with water and dried. The material thus obtained is then formulated for use as an adhesive composition, especially as a paper coating binder.

The *Micrococcus cerificans* bacteria cells are grown and harvested by supplying a mixture of 1 to 10 weight percent of a C_{11} to C_{30} paraffinic petroleum feed in an inorganic salt growth medium containing air (or other oxygen-containing gas) and water to a stirred reactor inoculated with *Micrococcus cerificans* and removing an effluent stream containing a mixture of bacteria, aqueous medium, and unconverted petroleum feed.

Then the bacteria cells are separated from the effluent stream, e.g., by centrifugation and dried. The reactor residence time usually ranges between 1 to 3 hours, and the amount of oxygen supplied usually ranges from 1.5 to 2.5 pounds per pound of product dried bacteria recovered per unit time.

Instead of employing a mixture of n-paraffins having from 11 to 30 carbon atoms, a single n-paraffin within this carbon number range can be used, e.g., n-hexadecane. The n-hexadecane used can be commercial n-hexadecane containing anywhere from 0.05 to 10.0 weight percent of n-olefins.

The *Micrococcus cerificans* cells can serve as a food or feed supplement for animals and/or humans so that a combined feed and adhesive process can be undertaken wherein a portion of the cells can be used for making adhesives and the remainder used for food.

In such a combined process, where some of the harvested cells are to be used as feed, it is preferable to exclude both polycyclic and monocyclic aromatic compounds from the hydrocarbon feed used to supply the carbon for bacterial cell growth.

Harvested cells can be recovered by centrifugation followed by drying (if desired), but neither centrifugation nor drying is required as the subsequent operations (lysing, basic extraction and isoelectric precipitation) can be carried out directly on a concentrated aqueous slurry of the *Micrococcus cerificans* cells, e.g., as obtained by allowing the aqueous product stream from the fermentation tank to settle in a settling tank to obtain a concentrated aqueous slurry having from about 2 to about 12 weight percent cells. However, any suitable recovery procedure can be employed to gain the harvested cells from the aqueous slurry in which they are grown. Examples of other methods of winning the harvested cells are filtration, settling, with or without filter aids or coagulants, etc. The cells need not be rendered nonviable prior to extraction of the protein therefrom.

The recovered cells are then subjected to a lysing operation. Lysing can be accomplished by grinding the cells or by subjecting a concentrated aqueous slurry of cells to heat or supersonic waves or by contact for several hours with aqueous acetone. Usually, however, the harvested *Micrococcus cerificans* cells are lysed by contact with aqueous acetone for time periods ranging from 1 to 5 hours, and usually 2 to 4 hours at ambient room temperatures, e.g., from 65° to 80°F.

The term "aqueous acetone" as used herein includes water-acetone mixtures or solutions containing from 85 to 95 weight percent acetone, the remainder being water (either tap water or distilled water).

The lysed cells are then extracted with an aqueous basic solution to remove the protein from the remainder of the cell. The basic extraction can be conducted using any suitable base, e.g., sodium hydroxide, potassium hydroxide, sodium sulfite, borax, ammonium acetate, etc.

The normality of the aqueous base solution can range from 0.05 to 5.0 N, usually ranges from 0.1 to 1.0 N, and preferably ranges from 0.1 to 0.5 N. The basic extraction can be accomplished at pH values ranging from 8 to 12 usually ranging from 9 to 11, and preferably at a pH of approximately 9.5 to 10.5.

The basic extraction is conducted at temperatures ranging from 20° to 40°C., usually from 22° to 35°C., and preferably from 25° to 29°C. for time periods of from 0.5 to 10 hours, usually 1 to 6 hours, and preferably from 2 to 4 hours. Then the extracted protein is subjected to centrifugation to remove insoluble debris present in the extract. The insoluble debris by-product contains 9 to 11% by weight nitrogen and can be used as by-product feed for animals.

After removal of the insoluble debris by centrifugation, filtration (or any other suitable removal technique), the extracted protein is isoelectrically precipitated by adjusting the pH of the basic extract to an acidic pH ranging from 5.2 to 3.2, usually from 4.5 to 3.8, and preferably to an acid pH of approximately 4.1 using any suitable acid. Suitable exemplary acids for isoelectric precipitation of the base extracted bacterial protein include, but are not limited to, the following: sulfuric acid, sulfurous acid, phosphoric acid, hydrochloric acid, etc.

The normality of the acid used for precipitation usually ranges from 2 to 10 N and preferably from 4 to 6 N, but an acid of any normality can be used to adjust the pH to the abovementioned level.

The precipitated protein curd is then collected by centrifugation, washed with water and then dried. Any convenient drying procedure can be employed, e.g., a drum-type dryer or spray dryer can be used.

Alternatively, the precipitated protein curd can be dried by slurrying in acetone followed by air drying at temperatures of 25° to 32°C. Yields of 35 to 55% of usable protein extract are uniformly realized (based upon dried cell weights). Of course, the exact yield of usable protein extract obtained will depend in part upon the protein content of the original harvested *Micrococcus cerificans* cells.

The material obtained by the abovementioned lysing, basic extraction, and isoelectric precipitation procedures is a high protein-containing material having a protein concentration ranging from 85 to 97 weight percent protein, usually 88 to 95 weight percent protein, and preferably contains 89+ weight percent pure protein.

A typical analysis (moisture-free basis) of the dried protein extract obtained in the manner set forth hereinabove from harvested *Micrococcus cerificans* bacteria cells is as follows:

C, 46 to 48%; H, 6.5 to 7.8%; N, 13.9 to 14.9%; ash, 1 to 5%; protein*, 87.93%; moisture, 9 to 10%; acidity, 2 to 3%

*The percent protein is calculated in the customary manner; viz, by multiplying the percent by weight of nitrogen (determined by the Kjeldahl method) by 6.31.

The physical properties of a typical sample of protein extracted from *Micrococcus cerificans* cells are tabulated on the following page.

Properties	Value
Total nitrogen (percent by weight)	14.4±0.5
Moisture content (percent by weight water)	9.5±0.5
Alkali requirement to solubilize protein extract (grams of sodium hydroxide per 100 grams of protein extract)	4.7±0.8
Free acidity (ml. of 0.1 N sodium hydroxide/gram to neutralize)	0.5±0.1

Example:

Component	Concentration (grams per liter)
n-Hexadecane	10.0
H_3PO_4	5.0
KCl	1.0
$CaCl_2$	0.5
$MgSO_4 \cdot 7H_2O$	0.2
$MnSO_4 \cdot 4H_2O$	0.2
$FeSO_4 \cdot 7H_2O$	0.2
NaCl	0.2
NH_4OH, amount added to maintain pH at 7.0.	

Micrococcus cerificans bacteria cells were grown on the above listed aqueous inorganic salt medium containing approximately 2 weight percent of n-hexadecane as the sole source of carbon. The harvested cells were harvested by means of centrifugation, filtered and dried at a temperature of 26°C. for 16 hours (under reduced pressure).

Then the cells were lysed by contact for several hours with an aqueous acetone solution containing 95 weight percent acetone with the remainder being water. These lysed cells were then dried at a temperature of 26°C. for 16 hours. The dry, lysed cells were then extracted for two hours with 0.15 normal sodium hydroxide aqueous solution. Sufficient sodium hydroxide was employed to maintain the pH above 10.0 with the actual pH range over the two hour period being from 10.0 to 11.0.

One third of these cells was retained for adhesive formulation after a single basic extraction as noted above. The remaining two thirds of the cells were again subjected to the same alkali extraction thus yielding doubly extracted cells. Following this the liquid protein extracts, including the singly and doubly extracted samples, were subjected to identical acidification using dilute 5 normal sulfuric acid to a pH of 4.1 to isoelectrically precipitate the extracted protein.

The precipitated protein curd was then collected by centrifugation, washed with water and dried. The yield of bacterial protein extract based on dried *Micrococcus cerificans* cells harvested was 45%. The material obtained by this procedure analyzed as 90% pure protein.

COMPANY INDEX

The company names listed below are given exactly as they appear in the patents, despite name changes, mergers and acquisitions which have, at times, resulted in the revision of a company name.

Allied Chemical Corp. - 45
Asahi Kasei Kogyo K.K. - 24, 61, 81
Atlantic Richfield Co. - 59, 113
British Petroleum Co. Ltd. - 4, 7, 12, 24, 86, 96, 132, 161, 170, 178, 183, 188, 190, 193, 196, 199, 209
Commercial Solvents Corp. - 21
Deutsche Akademie der Wissenschaften zu Berlin - 177, 182
Esso Research & Engineering Co. 26, 32, 64, 73, 102, 145, 159, 165, 206, 211
Gulf Research & Devel. Co. - 129
Institut Francais du Petrole des Carburants et Lubrifiants - 18, 141, 173
Institute of Gas Technology - 47
Inventa AG - 37
Kyowa Hakko Kogyo Co., Ltd. - 42, 49, 52, 157, 202
Mobil Oil Corp. - 68, 83, 91, 93
Nalco Chemical Co. - 121
Phillips Petroleum Co. - 29, 55, 98, 109, 127
Socony Mobil Oil Co., Inc. - 80
Stamicarbon NV - 39
Sun Oil Co. - 134
Takeda Chemical Industries, Ltd. - 14
Texaco Development Corp. - 76
U.S. Secretary of the Interior - 34

INVENTOR INDEX

Alquist, H.E., 29
Andreen, B., 47
Aoki, Y., 156
Avrillon, R., 18
Banno, I., 14
Behrens, U., 176, 182
Belavsky, M.A., 71
Belenky, S.I., 71
Belova, L.A., 71
Bennett, I.C., 196
Bernheimer, R., 159
Bonavita, J.B., 170
Cahn, R.P., 165
Carosella, J.M., 34
Cavallo, J.J., 59
Champagnat, A., 4, 7, 12, 167, 199, 209
Chepigo, S.V., 71
Clark, A.J., 193
Corrick, J.D., 34
Coty, V.F., 91
Dasinger, B.L., 206
Deguchi, T., 202
de Mayo, C., 12
Douros, J.D., Jr., 26, 32, 64, 73
du Chaffaut, J.A., 196
Filosa, J.A., 4, 7, 96, 132, 183, 188
Frankenfeld, J.W., 165, 206, 211
Frankowlak, S., 18
Friedman, M.R., 159
Furlong, L.E., 145
Furuya, A., 156
Gabert, A., 176, 182
Gatellier, C., 18, 141, 173
Glazman, B.A., 71
Glikmans, G., 18, 141, 173
Gorring, R.L., 83
Goux, R., 190
Griehl, W., 37
Guenther, K.R., 102
Harris, R.G., 76
Heilweil, I.J., 68
Herndon, J.F., 64
Hines, P.R., 59
Hitzman, D.O., 29, 127
Hondermarck, J.C., 190
Hoshiai, K., 49, 52
Hosler, P., 134
Hunt, R.J., 193
Iguchi, T., 24, 61, 81
Imada, O., 49, 52
Jiricka, J., 182
Joris, G.G., 45
Kimura, K., 42
Korotaev, O.G., 71
Kuno, M., 14
Lahl, W.J., 32
Lane, B.M., 24, 86, 96, 178, 190, 199, 209

Leathen, W.W., 129
Leavitt, R.I., 68, 93
Mako, P.F., 113
Maximenko, N.S., 71
McCoy, C.J., 73, 211
Munk, V., 182
Nakano, T., 61, 81
Nakao, Y., 14
Naslund, L.A., 32, 73
Ohshima, K., 42
Oppermann, R.A., 121
Orgel, G., 45
Perkins, M.B., 64, 102, 145
Phillips, U.A., 21
Pietrusza, E.W., 45
Pohland, D., 176, 182
Raymond, R.L., 80
Ringpfeil, M., 177, 182
Rodionova, G.S., 71
Rose, M.J., 34
Rudel, M., 176, 182
Samejima, H., 202
Sattler, K., 177, 182
Somov, V.S., 71
Stevens, N.J., 64
Suter-Homuth, C., 37
Sutton, J.A., 34
Takayanagi, M., 157
Takeda, I., 24, 61, 81
Tanaka, K., 42
Tanaka, M., 49, 52
Teranishi, H., 202
Tsuzuki, K., 61, 81
Vernet, C.L., 86
Vorobiova, G.I., 71
Voronkov, V.G., 71
Wegner, E.H., 55, 98, 109
Wen, C.-Y., 47
Woldendorp, J.W., 39
Wolnak, B., 47
Yamamoto, M., 42
Zayhekian, P., 165

U.S. PATENT NUMBER INDEX

2,769,750 - 76
3,186,922 - 161
3,193,390 - 4
3,224,946 - 80
3,257,289 - 12
3,264,196 - 132
3,268,412 - 199
3,268,413 - 86
3,268,414 - 24
3,268,419 - 209
3,271,266 - 96
3,293,145 - 68
3,308,035 - 26
3,337,413 - 55
3,347,688 - 211
3,355,296 - 145
3,384,491 - 102
3,411,989 - 14
3,414,477 - 73
3,418,208 - 91
3,427,223 - 165
3,474,001 - 93
3,489,648 - 109
3,508,927 - 64
3,510,401 - 98
3,510,403 - 86
3,520,777 - 183
3,522,147 - 188
3,523,064 - 37
3,525,671 - 127
3,530,039 - 159
3,536,585 - 178
3,540,983 - 34
3,546,071 - 32
3,560,341 - 7
3,578,566 - 177
3,585,179 - 202
3,586,605 - 134
3,591,455 - 121
3,594,277 - 113
3,595,749 - 193
3,616,209 - 190
3,616,214 - 24
3,616,216 - 98
3,616,248 - 182
3,616,249 - 59
3,619,368 - 39
3,620,927 - 129
3,622,465 - 45
3,625,832 - 173
3,627,637 - 21
3,630,843 - 157
3,634,194 - 206
3,635,796 - 49
3,639,210 - 42
3,642,577 - 83
3,642,578 - 29
3,645,846 - 52
3,649,459 - 47
3,654,084 - 71
3,655,511 - 170
3,658,647 - 81
3,674,640 - 61
3,698,998 - 18
3,701,713 - 196
3,702,283 - 141

NOTICE

Nothing contained in this Review shall be construed to constitute a permission or recommendation to practice any invention covered by any patent without a license from the patent owners. Further, neither the author nor the publisher assumes any liability with respect to the use of, or for damages resulting from the use of, any information, apparatus, method or process described in this Review.

Complete copies of the patents described in this book may be obtained at a cost of $0.50 per patent prepaid. Address order to the Commissioner of Patents, U.S. Patent Office, Washington, D.C. 20231.

FOOD AND BEVERAGE PROCESSING INDUSTRIES 1971

This book will serve as a useful guide to the United States food processing industry. Used properly, the book will become a valuable help to you in your sales and acquisition programs with the food companies included. There are some 3,500 manufacturing firms in this book, a most comprehensive listing.

The usefulness of this book derives from its organization. It is divided into two sections. The first section is an alphabetical listing by name of the approximately 3,500 United States food firms giving, in addition to the correct name of the company, both its address and zip code. Also given is concise, pertinent information (where available) such as:

(a) Annual Sales
(b) Number of Employees
(c) Name of Chief Executive
(d) Product Types

The second section is arranged numerically according to zip code with the companies once again listed alphabetically within their proper zip code numbers, thereby providing you with an easy-to-use, invaluable geographical index to the food industry of the United States.

The information provided by this book offers a valuable and easily used tool for locating the primary buying power in the food industry—a major factor in good sales planning.

This book therefore offers you:

(1) The most complete and up-to-date list of food processing companies available today, including basic information on each company (where available).
(2) An invaluable mailing list, not only by virtue of the alphabetical company name list, but also by the numerical zip code list.
(3) A help to you in your marketing efforts by providing you with the means for forward geographical planning—the maximum number of personal visits and calls per day for any particular area your salesman finds himself in, is easily determined by reference to the zip code listing.
(4) A valuable guide to planning acquisitions. The geographical section is particularly useful in that you can pinpoint food companies in specific areas of the country.
(5) Employment guide to this industry.

By using this guide in your marketing and acquisition planning, you can achieve the maximum return on your time and manpower outlay.

169 pages.

PREPARED SNACK FOODS 1973

by M. Gutcho

Food Technology Review No. 2

The April 24, 1972 issue of the "Chemical Marketing Reporter" places the value of the U.S. snack industry at $2 billion a year. It is therefore not surprising that many companies are working on improving existing processes or on developing new snack foods altogether.

The dictionary defines a snack as "food eaten between meals," which implies that snacks purchased in a supermarket or other food store should already be prepared, ready to eat. The present book therefore deals to a large extent with what is called "crisps" in the food processing industry, i.e., potato chips, corn chips, popcorn, expanded products, nuts and nut products, and french fried potatoes. The preparation of additional crisps, such as onion rings and apple crisps, is described under miscellaneous snack products.

151 patent-based processes. A partial and condensed table of contents follows. Numbers in ( ) indicate the number of patents per topic. Chapter headings are given, followed by examples of important subtitles.

1. POTATO CHIPS (39)
Prepared from Raw Potato Slices
Immersion in Hot Fat under Vacuum
Radiant Heat Cooking
Microwave Heating
Coating with Amylose
Low Fat Chips
Cooking in Fluidized Bed
Adding Methionine to Oil
3-Methylthiopropylamine Hydrochloride as Antioxidant
Preparation from Dough Bases Containing Fresh Potato Puree
Addition of Stearyl Glyceridyl Citrate
Dough Bases Containing Dried Potato Products
Gluten + Gelatinized Amylopectin
Puffing under Infrared Radiation
Structure-Forming Means
Packaging of Chip Type Snacks
Low Fat Puffed Products
Potato-Coated Snacks
Imparting Fried Taste and Appearance to Baked Goods

2. CORN CHIPS (9)
Addition of Corn Starch, Salt, and Carboxymethylcellulose to Gelatinized Mass
Rolling and Sheeting Corn Dough
Shaping of Corn Dough

3. POPCORN (8)
Coloring and Flavoring Popcorn
Packaging Popcorn
Popcorn Products

4. EXPANDED PRODUCTS (34)
Controlled Expansion
Extrusion
High Protein Snacks
Low Calorie Snacks
Agglomerated Cereal-Legume Mixture
Vacuum Puffing
Microwave Puffing
Fluidizing Gas Puffing
Deep Fat Fried Products
Spun Protein-Starch Fibers
Filled Puffed Products
Shaped & Patterned Snacks

5. NUTS AND NUT PRODUCTS (22)
Partially Defatted Peanuts
Retaining Peanut Skins During Roasting
Coating & Coloring Nuts
Edible Soybean "Nuts"
Synthetic Nutmeats
Fused Casein + Soybean Fibers

6. FRENCH FRIED POTATOES (24)
From Raw Potatoes
Treatment with Cross-Linking Agents
Surface Treatments in General
Dehydration Methods
Control of Moisture Content
From Potato Dough Bases
Binders & Agglomerates
Removal & Recovery of Excess Oil

7. MISCELLANEOUS SNACKS (15)
High Protein Chips
Bacon Flavored Chips etc.
Fruit-Based Snacks
Convoluted Apple Crisps
Banana Chips
Batter-Coated Onion Rings
Onion-Flavored Farinaceous Rings
Improving Characteristics of Pretzels and Croutons

ISBN 0-8155-0474-8 **289 pages**

EDIBLE COATINGS AND SOLUBLE PACKAGING 1973

by R. Daniels

Food Technology Review No. 3

Edible and soluble packaging can be used to

- — retard moisture transmission
- — serve as a barrier to oxygen or other gases
- — prevent loss of volatile components and flavor
- — resist transfer of oils and grease
- — serve as a barrier to bacterial invasion
- — provide protection from mechanical injury — or at least minimize crumbling and breakage
- — isolate reactive ingredients from each other.

The processes in this book are grouped according to the food products to which the edible or soluble coatings are applied. The subjects of edible and soluble capsules for medicines and of casings for sausages are so vast that they are treated in separate volumes.

118 patent-based processes. A partial and condensed table of contents follows. Numbers in ( ) indicate the number of patents per topic. Chapter headings are given, followed by examples of important subtitles.

1. **AMYLOSE, CELLULOSE, COLLAGEN, AND PROTEINS (17)**
   Amylose Coagulation from Caustic Solutions
   Sulfate Coagulation Methods
   Dispersible Amylose — Chemical Treatment Autoclaving
   Enzyme-Modified Collagen Films
   Cellulose Coatings
   Mineral Oil Compositions
   Plasticized Ethyl Cellulose
   General Mills Protein Films
   Soluble Protein Container

2. **COATINGS FOR FRUIT AND VEGETABLES (8)**
   Lecithin Emulsions
   Vinyl Acetate Polymers
   Gloss Coatings for Fruit

3. **RAISINS, DATES, AND OTHER STICKY FOODS (14)**
   USDA Processes for Raisins
   Controlling Coating Thickness
   Amylose-Protein Laminates
   Coating Emulsions for Confections

4. **NUTS AND NUT PRODUCTS (8)**
   One Step Coloring and Sealing

5. **MEAT AND FISH COATINGS (15)**
   Gelatin Coatings
   Alginate Coatings
   Water-in-Oil Emulsions
   Permeable Films
   Specialty Coatings for Fish

6. **CEREAL TYPE PRODUCTS (9)**
   Silane Surface Treatment
   Candy-Coated Cereal-Nut Mixture
   Dry Instant Rice Pudding

7. **BAKERY PRODUCTS (13)**
   Pillsbury Coatings
   Protective Glazes for Doughnuts
   Edible Container for Hot Foods

8. **ENCAPSULATION (12)**
   National Cash Register Processes
   Southwest Research Process
   Protein-Based Encapsulation
   Dextrin Encapsulation
   Wheat Gluten Base
   Gelatin Capsules
   Fat Encapsulation
   Dextrins for Acetal Fixation

9. **COATINGS FOR SALT-TYPE INGREDIENTS AND VITAMINS (9)**
   Salt Coating Processes
   Rutgers Fat Coating Method
   Cellulose Coating Methods
   Prolamin Coating
   Coating with Molten Curing Salt
   Emulsion Coatings
   Gelatin Emulsion Method
   Saturated Fatty Acids
   Aqueous Amylose Emulsions

10. **MISCELLANEOUS AND SPECIALTY COATINGS (13)**
    Coatings Containing Sorbic Acid
    Air-Dried Emulsion Coating
    Hardened Shortening
    Coated Fibrous Proteins
    Coatings to Prevent Caking
    Fat and Starch Coated Feedstuffs
    Coating with Immiscible Powders
    Release Coatings
    Batter Coatings
    Alginate Gel Coating
    Coatings for Extruded Food Products
    Gelatin Membrane Separator

ISBN 0-8155-0475-6 **360 pages**

TEXTURED FOODS AND ALLIED PRODUCTS 1973

by M. Gutcho

Food Technology Review No. 1

This book deals primarily with the preparation of meat-like foods from readily available protein, such as soy protein, peanut protein, and wheat gluten. Techniques employed by the textile and plastics industries have been borrowed and applied to plant protein.

Processes have been developed for spinning protein into odorless, tasteless filaments, which provide the necessary texture for simulated food products, or as meat extenders. Fibrous textures can also be obtained by nonspinning techniques, such as thermoplastic extrusion.

This new category of food has been designated by the Food and Drug Administration as "textured protein products." These sources of protein, when used as meat, in order to be acceptable to the consumer, must simulate the texture, chewiness, flavor, and appearance of a natural meat product.

Textured plant protein has gained acceptance as a meat extender, especially in school lunch and geriatric programs, which permit a 30% addition of plant protein to meat items. The book reviews 78 U.S. patents, most of them issued quite recently. Numbers in ( ) indicate a plurality of processes per topic. Chapter headings are given, followed by examples of important subtitles.

1. PREPARATION OF EDIBLE SPUN PROTEIN FIBERS (24)
- Boyer Processes
  - Oriented Edible Protein Fibers made by Stretching Filaments
  - Modifying Agent Added Prior to Spinning
  - Controlling Fiber Toughness
- Spinning Solutions
  - Protein-Alginate Compositions
  - Casein & Soybean Meal Spinning Dope
  - Polymeric Carbohydrate Gel
- Coagulating Bath Modifications
  - Alkali Metal and Aluminum Salts
  - Advantages of Sulfites
- Binders for Spun Fibers
- Impregnating and Compressing Tows
- Meat-Like Products from Fibers
- Spun Fibers in Sausages

2. VARIOUS TECHNIQUES FOR MEAT-LIKE TEXTURED FOODS (19)
- From Chewy Protein Gel
- Expanding the Cellular Structure
  - Microwave Energy
  - Cysteine and Hypophosphites
  - Defatted Soy Meal + Sulfur Compounds
  - Fat + α-Cellulose
  - Bacon-Like Product from Oilseed Protein
- Obtaining Fibers Without Spinning
  - Addition of Phosphate Ions
- Miscellaneous Processes

3. SIMULATED MEAT PRODUCTS (10)
- Wheat Gluten Base Products
  - Dehydrated Gum Gluten
  - Vitalized Wheat Gluten
  - Defatted Soy Flour Granules
- Spreads and Luncheon Meats
  - From Soy or Peanut Protein Flakes plus Aqueous Emulsions of Fat
  - Gel-Like Dispersions of Edible Oils Within Albumin Film Formers
  - Simulated Meat Baby Food from Unstretched, Unoriented, Coagulated Oilseed Protein plus Proteolytic Enzymes
- Other Products
  - Artificial Adipose Tissue Utilizing Alginate Gel Matrix for Fat
- Meat-Like Chunks from Soybean Meal
  - Pieces Impregnated with Coconut Oil
  - Multicolored Simulated Chicken Dye Injected into Extruder Mixer

4. OTHER TEXTURED SIMULATED FOODS (12)
- Nut Meats
  - Fused Casein or Soybean Fibers
- Snack Foods
  - Bacon-Flavored Chips from Farinaceous Dough + Texturizing Agent
  - Puffed Snacks from Protein + Starch Fibers
- Cereals
  - Encased in Spun Sugar Filaments
- Other Products
  - Simulated Meringue
  - Synthetic Granular Caviar

5. COMPATIBLE FLAVORS (13)
- Protein Hydrolyzates + 5'-Nucleotides
  - Yeast Autolyzate
  - Sulfur-Containing Polypeptides

ISBN 0-8155-0463-2 **315 pages**

COCOA AND CHOCOLATE PROCESSING 1972

by H. Wieland

Food Processing Review No. 27

Of all the world's countries, the USA is the largest consumer of cocoa products, equivalent to about 20% of the world production of raw beans. According to a U.S. Dept. of Commerce report, the 1971 U.S. consumption of confectionery products containing chocolate totaled 4,000 million av. pounds with a sales value of $2,000 million.

Partially defatted cocoa products are more palatable to most people, but manufacturers are finding new ways to decrease the amount of natural cocoa bean products in finished goods in order to reduce costs. Synthetic flavors, flavor enhancers, cocoa butter substitutes from vegetable oils, emulsions and water extracts are included in the processes and formulations given here, so as to allow reduction of natural raw materials which must be imported from tropical countries.

104 patent-based processes mostly since 1960. Numbers in ( ) indicate number of processes per chapter. Chapter headings are given here, followed by examples of important subtitles.

1. **ROASTING, MILLING, CONCHING (10)**
   - Continuous Process for Cocoa Nibs
   - Jet Mill Grinding
   - Pressure Roasting
   - Continuous Conching Process
   - Film or Liquid Conche Process
   - Pilz Spray Conche Process
2. **PLAIN AND MILK CHOCOLATE (10)**
   - Programmed Milling
   - Incorporating Nuts and Fruits
   - Chips with High Softening Point
   - Low Fat Chips
   - Chips for Ice Cream
   - Continuous Fudge Process
   - White Chocolates
3. **COATINGS AND LIQUID-FILLED GOODS (11)**
   - Thermostable Coatings
   - Stable Polymorphic State by Homogenization
   - Uniformly Tempered Chocolate
   - Reduced Viscosity with Lecithin
   - Non-Leaky Liquid Centers
   - Method for Filling Liquid Centers
4. **TRIGLYCERIDES AS HARD BUTTERS (16)**
   - Compatible Triglyceride Compositions
   - Stable Bloom Resistance
   - Tropical Chocolate Coatings
   - Fractions from Shea Butter
   - Anti-Bloom Agents
   - Conversion of Cocoa Butter to Free Acids
5. **INTERESTERIFIED TRIGLYCERIDES (10)**
   - Rearranged Hard Butters
   - High Yield Randomization Process
   - Seiden Corandomized Butters
   - Hydrogenated Palm Kernel Oil
   - Tripalmitin and Tristearin
   - Emulsifiers for Coatings
6. **FLAVORS AND FLAVOR ENHANCERS (14)**
   - Base Flavor from Protein Hydrolyzate
   - Recovery of Aroma
   - Flavoring from Cocoa Shells
   - Ethyl Maltol Enhancer
   - Pyrazine Enhancer
   - Alkylthiazolidine Enhancer
7. **CHOCOLATE DRINK POWDERS (12)**
   - Dispersible Cocoa Powder
   - Dispersion with Wetting Agents
   - Lecithin for Agglomeration
   - Cocoa Bases for Chocolate Drinks
   - Defatted Cocoa Base
   - Agglomerated Cocoa with Fat
   - Binding with Emulsifiers
   - High Fat Milk Powders
   - Wettable Cocoa-Sugar Mix
   - Flat Spray Moisture Process
   - Roto-Dry Process
8. **CHOCOLATE BEVERAGES AND SYRUPS (9)**
   - Concentrates for Chocolate Drinks
   - Dutched Cocoa for Milk Drink
   - Fat-Free Cocoa Extracts
   - Stabilized Chocolate Beverage
   - Carragheenate Stabilizers
   - Soy Milk Chocolate Drink
   - Cocoa-Flavored Oils
   - Chocolate Water-in-Oil Emulsions
9. **OTHER CHOCOLATE PRODUCTS (12)**
   - Soft Plastic Cooking Chocolate
   - Chocolate Marshmallow
   - Chocolate Water-Icings
   - Foam Chocolate Topping
   - Ready-to-Spread Frosting
   - Milk Crumb Intermediate Product
   - Vacuum-Expanded Malted Milk Balls
   - Chocolate-Flavored Banana Powder
   - Improved Chocolate for Ice Cream
   - Chocolate Granules
   - Cocoa Butter Binder for Tablets
   - Cocoa Butter Suppositories

300 pages

FOOD CANNING TECHNIQUES 1972

by M. Gutterson

Food Processing Review No. 26

Canning, as opposed to frozen food preservation, makes use of metal and glass containers almost exclusively and relies on heat processing of food in hermetically sealed containers. Much research has been done toward flavor preservation and elimination of harmful microorganisms, making long term nonrefrigerated storage a present-day reality.

107 Patent-based pocesses. A partial table of contents is given here. Numbers in ( ) indicate numbers of processes per subject.

1. GENERAL TECHNIQUES
- CANNING SYSTEMS (8)
  - High Temperature Radiation
  - Heating in Fluidized Bed
  - Hydrostatic Cooking
  - Radio Frequency Sterilization
  - Microwave Sterilization
- ASEPTIC CANNING (3)
  - Chemical Bactericides
  - Superheated Steam
  - Infrared Radiation
- CLOSURES (4)
  - Thermoplastic Caps
  - Easily Opened Containers
  - Vacuum Seals
  - Reclosable Seals
- CAN COATINGS (4)
  - Butadiene + α -Methylstyrene
  - Palladium for O_2 Removal
- DEOXYGENATION (4)
  - Oxygen-Permeable Barriers
  - Bisulfites
  - Water Synthesis Catalyst
  - Enzymatic Oxygen Removal
- CONTAINERS (3)
- CHEMICAL PRESERVATION (4)
  - Alkyl p-Hydroxybenzoates
  - Benzohydroxamic Acid
  - Hydrochloric Acid

2. MEAT AND POULTRY
- WHOLE PIECES & CHUNKS (7)
  - Eliminating Air Pockets
  - Inserting Meat into Cans
- COMMINUTED MEAT (6)
  - Use of Ascorbic Acid
  - Plus Sodium Nitrate
  - Alginate and Mg Salts
- SMOKE & SMOKE FLAVORS (2)
- MEAT RELEASE AGENTS (1)
- SEMICOOKED MEAT (1)
- BARBECUED RIBS (1)
- POULTRY (2)
  - Breaded Fried Chicken

3. SEAFOOD
- PREPARATION OF FISH (4)
  - Degassing
  - Storage under CO_2
- PRECOOKING (3)
- FISH CANNING SYSTEMS (4)
- FISH PRODUCTS (3)

4. VEGETABLES, FRUITS, JUICES
- CANNING SYSTEMS (2)
- TOMATOES (4)
  - Aseptic Filling
  - Acid Treatment
- CITRUS PRODUCTS (4)
  - Low Pressure Treatment
  - Flavor Improvement
- APPLES (1)
- CREAMED CORN (1)
- PICKLES (4)
- PEAS—Color Preservation (1)
- OLIVES (2)

5. DAIRY PRODUCTS
- MILK (8)
  - Prevention of Thickening
  - Elimination of Cooked Flavor
  - Viscosity Control Agents
  - Infants' Formulations
  - Condensed Cream (2)
- CHEESES (2)

6. BEVERAGES
- BEER AND ALE (2)
- SOFT DRINKS (2)
  - Air-Free Packaging
  - Prevention of Can Corrosion
- CANNING WATER (1)
  - Retaining Freshness

7. DOUGHS & BAKED GOODS (4)

8. VARIOUS PRODUCTS (5)
- Aerosol Containers

238 pages

SAUSAGE PROCESSING 1972

by Dr. E. Karmas

Food Processing Review No. 24

The detailed, descriptive process information included in this Food Processing Review is based on U.S. patents relating to sausage production. The material involving 136 processes, developed since 1960, has been divided into two parts—Emulsion Ingredients and Processing Methods. The first part emphasizes various emulsion additives from the viewpoint of cured color, flavor, emulsion stability, and ingredient control. The second part deals with sausage processes such as stuffing, linking, cooking, and peeling; as well as the production of conventional and some novel sausage products.

Partial Table of Contents showing process examples:

1. ACCELERATED DEVELOPMENT AND STABILIZATION OF CURED COLOR (18)
    - Glucono Delta Lactone
    - Methyl 2-Ketogluconate
    - Soluble Calcium Salts
    - 2,3-Dihydroxy-1-cyclohexenone
    - 1,2,3-Cyclohexanetrione
    - Nitric Oxide Gas
    - Dialuric Acid
    - Fumaric Acid
    - Ethylenediaminetetraacetic Acid
    - Acid Phosphates
    - Citrates and Ascorbates
    - Sorbitol
2. FLAVORING & COLORING ADDITIVES (3)
3. PREVENTION OF RANCIDITY (4)
    - Hot Carcass Processing
    - Oxygen-Impermeable Casings
    - Ground Mustard Seed
    - NaCl Coated with Antioxidants
4. FERMENTED FLAVORS (3)
5. TEXTURIZING AGENTS (13)
    - Phosphate Gels
    - Calcium Caseinate
    - Methyl Cellulose
    - Wheat Flour-Base Binder
    - Milk Solids with Enzyme
    - Bovine Blood Plasma
    - Salt-Soluble Protein
6. EMULSION STABILIZERS (5)
    - Magnesium Ions
    - Polymeric Phosphates
    - Other Phosphates
7. LOW TEMPERATURE DEFATTING OF MEAT (3)
8. LOW TEMPERATURE DEFATTING OF ADIPOSE TISSUE (3)
9. FATTY TISSUE RENDERING (2)
10. MEAT RECOVERY FROM BONES (4)
11. COMPOSITION CONTROL (2)
12. PRODUCT AND INGREDIENT IDENTIFICATION (2)
13. STUFFING OPERATIONS (10)
    - Deaerated Emulsion
    - Controlled Stuffing Rate
    - Continuous Stuffing
    - Metered Stuffing
14. LINKING AND TYING (13)
15. FORMING AND SHAPING (3)
    - Calcium Alginate Coating
    - Surface Treatment in Acid
16. MISCELLANEOUS METHODS (4)
    - Smoking Meat While Comminuting
    - Pumping Meat Emulsion
17. HOT AIR COOKING (6)
    - Static Air Heating
    - Control of Humidity
18. COOKING IN LIQUID (3)
19. COOKING WITH HEAT EXCHANGER
20. ELECTRICAL COOKING (14)
    - Forming and Cooking
    - Forming Devices
    - Skin-Forming and Dyeing
    - Heating of Flowing Emulsion
21. RELEASING THE CASING (2)
22. REMOVING THE CASING (3)
23. FRESH PORK SAUSAGE (3)
24. DRY SAUSAGE (8)
    - Buffered Humidity Zone
    - Cellulose Casings
    - Reduction of Drying Time
25. VARIOUS PRODUCTS (2)
    - Continuous Method of Making Small Smoked Sausages
    - Liver Sausage in Laminate Casing
26. NOVELTY PRODUCTS (2)
    - Annular Sausages
    - Multicolored Sausages

218 pages

ENZYMES IN FOOD PROCESSING AND PRODUCTS 1972

by H. Wieland

Food Processing Review No. 23

In recent years a vast engineering and technological break-through was achieved to meet the large demands for detergent enzymes. Experience has been obtained in large-scale and continuous fermentation and isolation and purification on a production basis. Because of this, enzymes applicable to food processing and products have become more available by technological transfer, and the alert food processor is urged not to miss this opportunity to improve his products in many ways. 101 Processes. Table of Contents follows. The numbers in parentheses indicate the number of processes allocated to each topic.

1. FRUIT AND VEGETABLE PROCESSING (13)
- Extraction of Vegetables
- Rapid Process for Fruit Puree
- Citrus Juices (4)
- Dehydrated Sweet Potatoes
- Enzymes for Candied Fruits
- Garlic Process
- Enzymes for Pickles
- Enzyme Pretreatment
- Enzymocidal Gas Treatment
- Low Sugar Pectins

2. STARCH AND SUGAR CONVERSION (9)
- Starch Saccharification (2)
- Starch to Maltose
- Dextrose to Levulose
- Gum Confections (2)
- Dietary Sugars from Starch
- Improved Specificity with MgO
- Low pH Dextrose

3. BAKED GOODS APPLICATIONS (12)
- Enzyme Treatment of Flour
- Bread Doughs (5)
- Stale Bread Prevention with Enzymes (2)
- Improved Bread Flavor
- Pie Dough
- High Protein Crackers
- Ready-to-Spread Frosting

4. CHEESE MAKING (11)
- Milk-Clotting Enzyme Product
- Reduced Curing Time
- Direct Acid Processes (2)
- Microbial Rennets (2)
- Enzymes as Rennet Substitutes
- Ripening Enzymes in Wrappers
- Reworked Cheese
- Low Fat Cheese
- Lipolyzed Milk Fat Products

5. MEAT TENDERIZATION (18)
- Preslaughter Treatments (3)
- Post-Rigor Mortis Treatment
- Uniform Tenderization (2)
- Pancreatic & Mesenteric Enzymes
- Activation of Enzymes in Meat
- Inactivation of Enzymes in Meat
- Pyrophosphates and Enzymes (2)
- Upgrading Quality
- Corned Meat Process
- Papain plus Bromelain
- Enzymes and Freezedrying
- Cold Water Buffered Enzyme
- Enzyme and Antibiotic Synergism
- Aerosol Enzyme Compositions

6. SPECIAL APPLICATIONS (13)
- Reducing Shrinkage
- Removal of Meat from Bones
- Fish Protein Concentrates (2)
- Collagen Tenderization
- Collagen Processing
- Cereal Processing (2)
- Nonsticky Pasta
- Lipoxidases from Soybean Whey
- Meatlike Proteins (2)
- Oyster and Shrimp Soup Mixes

7. FLAVORS BY MEANS OF ENZYMES (13)
- Flavor Producing Enzymes
- Flavor-Improved Dried Fruit
- Improved Dairy Flavors (3)
- Soy Sauce and Soy Beans (2)
- Garlic & Onion Flavor Control
- Cocoa Flavoring
- Legumes and Mushrooms (2)
- Citrus Oils
- Nucleotides from Nucleic Acids

8. DEOXYGENATING AND DESUGARING (6)
- Glucose Oxidase
- Desugaring Egg Whites
- Improving Color of Egg Products
- Haze-Free Low Calorie Beer
- Polymer + Enzyme for Chillproofing
- Removing Pectins from Agar

9. ENZYME STABILIZATION (6)
- Stabilization by Coating
- Modified Water Insoluble Enzymes
- Preservation of Tomato Pectase
- Stabilization of Diastase
- Stabilized Papain Solutions
- Stable Glucose Oxidase-Catalase

260 pages

SEAFOOD PROCESSING 1971

by M. Gillies

Food Processing Review No. 22

The consumption of luxury seafood—crab, shrimp, lobster, oyster—increases 25% a year, but prices of these delicacies rise also. A large part of the expense comes from the hand labor traditionally involved in shelling or shucking the seafood. Many processes in this book deal with these problems and have proved their usefulness in commercial application. Probably the most interesting development is fish protein concentrates which have largely overcome the disadvantages of fresh-caught fish.

Includes ca. 80 processes developed since 1960. Partial contents:

I. PRESERVATION (13 processes)
- 1. ON FISHING VESSELS (2)
- 2. CHEMICAL METHODS (6)
  - 5-Aminohexahydropyrimidine
  - Sorbic Acid and Antioxidant
  - 2-Haloacetoacetic Acids
  - Polyphosphate Dip
  - Chelates and N_2O (2)
- 3. EDIBLE COATINGS (2)
  - Proteinaceous Gel
  - Starch and Alginates
- 4. VARIOUS PRESERVATIVES (3)
  - Irradiation of Frozen Fish
  - Microwave and Gas
  - Formic Acid on Whale Meat

II. CANNING PROCESSES (11)
- 5. TUNA AND SIMILAR (7)
  - Cleaning and Precooking
  - At Lowered Pressures
  - Semipermeable Films
  - Phosphate Treatment
  - Cooling of Precooked Tuna
  - Thawing (2)
- 6. SARDINES (1)
- 7. FORESTALLING STRUVITE (3)
  - By Phytic Acid
  - With EDTA
  - Electrical Prevention

III. PROTEIN CONCENTRATES (12)
- 8. MECHANICAL MEANS (2)
  - Apparatus
  - Admixture of Antioxidants
- 9. CHEMICAL MEANS (2)
  - VioBin Processes
  - From Oily Fish
- 10. BIOLOGICAL MEANS (5)
  - Hydrolysis with Yeast
  - Aspergillus flavus
  - Speed Fermentation
  - Lactic Acid Bacteria
  - Surfactant and Bromelain
- 11. STICK WATER PROTEINS (3)
  - Peptizing Insolubles
  - 2-Stage Digestion
  - Fat Removal and Dehydration

IV. MOLLUSKS AND SHELLFISH (15)
- 12. SQUID (2)
  - Skinning and Filleting
  - Preparing Surumeika
- 13. BIVALVES (4)
  - Opening with Chemicals
  - Shucking Oysters
  - Treating "Fainted" Bivalves
  - Inedible to Edible Clams
- 14. CRUSTACEANS (9)
  - Removing Meat
  - From Tails Only
  - Chemical "Soft Shells" (2)
  - Freezing Lobsters
  - Removal of Hemocyanin
  - Shrimp Color Improvement
  - Cooking Shrimp
  - Peeling Shrimp
  - Preventing Curl

V. CONSUMER PRODUCTS (17)
  - Abalone Steaks
  - Fish Sticks
  - Other Shapes
  - Fish Chips
  - Clam Snacks
  - Fish-Potato Mixtures
  - Oyster Stew Mixes
  - Shrimp Soup Mixes
  - Flavors from Crustaceans
  - Batter-Coated Seafoods (4)

VI. ANIMAL FOODS (11)
  - From Fish Solubles (4)
  - Chick Growth Factors
  - "Trash Fish"
  - Grinding Dogfish
  - Pelletized Feeds
  - Mink Feed
  - Pet Food from Clam Waste

206 pages.

FRUIT PROCESSING 1971

by M. Gutterson

Food Processing Review No. 21

Aside from canning, pasteurizing, sterilizing, dehydrating, freezing, and freeze-drying, one prime objective for the modern fruit processor is the retention of the characteristics of freshly picked fruit. A further goal is to minimize the quantities of preservatives, when recourse to the use of chemical substances must be taken. Preventing the growth of undesirable microorganisms in fresh fruit products and thus insuring relatively long "showcase life" is the subject of many of the 140 processes contained in this book.

All processes have been selected with the intention of eliminating the requirement of costly processing equipment; notwithstanding the purpose of providing fruits and fruit products that are highly acceptable by organoleptic tests, and which can be considered as "slow spoilers" on the basis of unsightly appearance or off-flavor.

The partial Table of Contents indicates the wealth of hard-to-get information collected here. This volume reviews fruit processing technology mostly from U.S. patents since 1960. Numbers in parentheses indicate the number of processes in each chapter, and **examples** of the processes found in each chapter are listed.

1. GENERAL TECHNIQUES (34)
   - Inhibiting Discoloration of Cut Fruit
   - Heat Treatment plus Sorbates
   - Formic Acid Esters to Retard Ripening
   - Degreening and Ripening
   - Increasing Cellular Permeability
   - Dehydration and Rehydration
   - Acetylated Monoglycerides
   - Microwave Dehydration
2. TREATMENT OF POMES (23)
   - Blanching of Apples
   - Continuous Process for Applesauce
   - Control of Scald and Browning
   - Color Stabilization—Coatings
   - Instantly Rehydratable Applesauce
   - Combined Freezedrying and Airdrying
   - Coating Apple Segments
3. CITRUS FRUIT PROCESSING (18)
   - Candying Citrus Fruit
   - Purees and Marmalade Bases
   - Waste Citrus Pulp Use in Juices
   - Recovery of Citrus Oil by I.R. Radiation
   - Preservation Processes—U.V. Radiation
   - Maceration and Homogenization
   - Trichloroethane and Methylene Chloride
   - Prevention of Mold Growth
   - Improvements in Mold Growth Prevention
   - Processing of Citrus Peel
   - Dehydrated Citrus Peel Products
   - Pastes from Citrus Peel
   - Coloring Citrus Fruits with Carotenoids
   - Coloring with Naphthol Derivatives
4. PROCESSING TECHNIQUES FOR BERRIES (20)
   - Cranberry Sauce—Cranberry Relish
   - Strained and Jellied Cranberry Sauces
   - Treatment of Grape Pomace to Extract Desirable Flavor and Color Constituents
   - Dehydration of Blueberries
   - Freezedried Strawberries for Cereals
   - Partial Dehydration and Impregnation of Berries with Syrup Prior to Freezing
   - Destruction of Microorganisms
5. PROCESSING TECHNIQUES FOR DRUPES (16)
   - Preparation of Cherry Purees
   - Production of Maraschino Cherries
   - Removal of Sulfur Dioxide from Cherries
   - Enzymatic Treatment of Cherries for Rapid Candying
   - Prevention of Discoloration in Cherries by Treatment with EDTA
   - Modern Pitting Techniques
   - Preservation of Peaches After Harvesting
   - Prevention of Browning of Frozen Peach
   - Vacuum Dehydration of Peaches
   - Apricots with Sun-dried Characteristics
6. PREPARATION OF DRIED FRUITS (10)
   - Preservation of Raisins with Silicone
   - Coating Raisins with Beeswax
   - Comminuting Dried Prunes for Fruit Bars
   - Packing Rehydrated Fruit in Nonmetal Foil
   - Fig Pellets for Baked Goods
   - Deliberate Sugar Inversion in Dates for Improving Texture
7. OTHER FRUITS (17)
   - Accelerated Ripening of Bananas with Lauryl Alcohol
   - Gases for Ripening
   - Controlled Release of Ethylene from Molecular Sieves
   - Flavor Improvement of Dehydrated Bananas Using Enzymes from the Peel
   - Delaying Senescence in Pineapples
   - Preservation of Cantaloupes by Injection of Inert Gases
   - Freezing of Avocados by a Two-Stage Process
   - Preservation of Avocados by Freezing and Other Means

224 pages.